MONOGRAPHS ON
APPLIED PROBABILITY AND STATISTICS

General Editors

M.S. BARTLETT, F.R.S. *and* D.R. COX, F.R.S.

SOME BASIC THEORY
FOR STATISTICAL INFERENCE

Some Basic Theory for Statistical Inference

E.J.G. PITMAN
M.A. D.Sc. F.A.A.
Emeritus Professor of Mathematics
University of Tasmania

CRC Press
Taylor & Francis Group
Boca Raton London New York

CRC Press is an imprint of the
Taylor & Francis Group, an informa business

First published 1979 by CRC Press
Taylor & Francis Group
6000 Broken Sound Parkway NW, Suite 300
Boca Raton, FL 33487-2742

Reissued 2018 by CRC Press

Library of Congress Cataloging in Publication Data

Pitman, Edwin J. G.
Some basic theory for statistical inference.

(Monographs on applied probability and statistics)
Includes bibliographical references.
1. Mathematical statistics. I. Title.
QA276.P537 519.5 78-11921
ISBN 0-470-26554-X

A Library of Congress record exists under LC control number: 78011921

ISBN 13: 978-1-315-89767-7 (hbk)
ISBN 13: 978-1-351-07677-7 (ebk)

Visit the Taylor & Francis Web site at http://www.taylorandfrancis.com and the
CRC Press Web site at http://www.crcpress.com

CONTENTS

PREFACE

This book is largely based on work done in 1973 while I was a Senior Visiting Research Fellow, supported by the Science Research Council, in the Mathematics Department of Dundee University, and later while a visitor in the Department of Statistics at the University of Melbourne. In both institutions, and also at the 1975 Summer Research Institute of the Australian Mathematical Society, I gave a series of talks with the general title 'A New Look at Some Old Statistical Theory'. That title indicates fairly well my intentions when I started writing this book.

I was encouraged in my project by some remarks of Professor D.V. Lindley (1972) in his review of *The Theory of Statistical Inference* by S. Zacks:

> One point that does distress me about this book—and let me hasten to say that this is *not* the fault of the author—is the ugliness of some of the material and the drabness of most of it... The truth is that the mathematics of our subject has little beauty to it. Is it wrong to ask that a subject should be a delight for its own sake? I hope not. Is there no elegant proof of the consistency of maximum likelihood, or do we have to live with inelegant conditions?

I share Lindley's dissatisfaction with much statistical theory. This book is an attempt to present some of the basic mathematical results required for statistical inference with some elegance as well as precision, and at a level which will make it readable by most students of statistics. The topics treated are simply those that I have been able to do to my own satisfaction by this date.

I am grateful to those who, at Dundee, Melbourne, or Sydney, were presented with earlier versions, and who helped with their questions and criticisms. I am specially grateful to Professor E.J. Williams, with whom I have had many discussions, and who arranged for and supervised the typing; to Judith Adams and Judith Benney, who did most of the typing, and to Betty Laby, who drew the diagrams.

E. J. G. P.

CHAPTER ONE

BASIC PRINCIPLES OF THE THEORY OF INFERENCE THE LIKELIHOOD PRINCIPLE SUFFICIENT STATISTICS

In developing the theory of statistical inference, I find it helpful to bear in mind two considerations. *Firstly*, I take the view that the aim of the theory of inference is to provide a set of principles, which help the statistician *to assess the strength of the evidence* supplied by a trial or experiment for or against a hypothesis, or *to assess the reliability of an estimate* derived from the result of such a trial or experiment. In making such an assessment we may look at the results to be assessed from various points of view, and express ourselves in various ways. For example, we may think and speak in terms of repeated trials as for confidence limits or for significance tests, or we may consider the effect of various loss functions. Standard errors do give us some comprehension of reliability; but we may sometimes prefer to think in terms of prior and posterior distributions. All of these may be helpful, and none should be interdicted. The theory of inference is persuasive rather than coercive.

Secondly, statistics being essentially a branch of applied mathematics, we should be guided in our choice of principles and methods by the practical applications. All actual sample spaces are discrete, and all observable random variables have discrete distributions. The continuous distribution is a mathematical construction, suitable for mathematical treatment, but not practically observable. We develop our fundamental concepts, principles and methods, in the study of discrete distributions. In the case of a discrete sample space, it is easy to understand and appreciate the practical or experimental significance and value of conditional distributions, the likelihood principle, the principles of sufficiency and conditionality, and the method of maximum likelihood. These are then extended to more general distributions by means of suitable definitions and mathematical theorems.

Let us consider the likelihood principle. If the sample space is discrete, its points may be enumerated,

$$x_1, x_2, \ldots$$

Suppose that the probability of observing the point x_r is $f(x_r, \theta)$, and that θ is unknown. An experiment is performed, and the outcome is the point x of the sample space. All that the experiment tells us about θ is that an event has occurred, the probability of which is $f(x, \theta)$, which for a given x is a function of θ, the likelihood function.

Consider first the case where θ takes only two values θ_0, θ_1. To decide between θ_0 and θ_1, all that the experiment gives us is the pair of likelihoods $f(x, \theta_0), f(x, \theta_1)$. Suppose that θ is a random variable, taking the values θ_0, θ_1 with probabilities p_0, p_1, where $p_0 + p_1 = 1$. Given the observed value x, the conditional probabilities of θ_0, θ_1 are proportional to $p_0 f(x, \theta_0)$, $p_1 f(x, \theta_1)$. The conditional odds of θ_1 against θ_0 are

$$(p_1/p_0)[f(x, \theta_1)/f(x, \theta_0)].$$

The prior odds are p_1/p_0, and all that comes from the experiment is the likelihood ratio $f(x, \theta_1)/f(x, \theta_0)$.

If $f(x, \theta_1)/f(x, \theta_0) = \infty$, then $\theta = \theta_1$, if $f(x, \theta_1)/f(x, \theta_0) = 0$, then $\theta = \theta_0$. Let c be a positive number. Denote the set of x points for which $f(x, \theta_1)/f(x, \theta_0) = c$ by A. If $x_r \in A$, the θ_1 conditional probability of x_r given A is

$$\frac{f(x_r, \theta_1)}{\sum\limits_{x \in A} f(x, \theta_1)} = \frac{c f(x_r, \theta_0)}{\sum\limits_{x \in A} c f(x, \theta_0)} = \frac{f(x_r, \theta_0)}{\sum\limits_{x \in A} f(x, \theta_0)},$$

which is the θ_0 conditional probability of x_r given A. Hence the conditional distribution of any statistic T will be the same for $\theta = \theta_1$ as for $\theta = \theta_0$. Thus when we know $f(x, \theta_1)/f(x, \theta_0)$, knowledge of the value of T gives no additional help in deciding between θ_0 and θ_1. We express all this by saying that when θ takes only two values θ_0 and θ_1, $f(x, \theta_1)/f(x, \theta_0)$ is a *sufficient statistic* for the estimation of θ. In the ordinary, non-technical sense of the word, all the information supplied by the experiment is contained in the likelihood ratio $f(x, \theta_1)/f(x, \theta_0)$. When θ can take many values, all the information about θ given by the experiment is contained in the likelihood function, for from it we can recover any required likelihood ratio. This is the likelihood principle.

For a discrete sample space, where the observed sample has a non-zero probability, this seems sound; but when we come to a continuous distribution, we are dealing with an observation which has zero probability, and the principle seems not so intuitively appealing. However, the extension to the continuous distribution seems reasonable when we regard the continuous distribution as the limit of discrete distributions.

If the probability measure p_θ, on the sample space has density $f(x,\theta)$ with respect to a σ-finite measure μ, it is still true that when θ can take only two values θ_0, θ_1, all conditional probabilities given $f(x,\theta_1)/f(x,\theta_0)=c$ (positive and finite) are the same for $\theta=\theta_1$ as for $\theta=\theta_0$. The likelihood ratio $f(x,\theta_1)/f(x,\theta_0)$ is a sufficient statistic; but it should be remembered that when the probability of the conditioning event is zero, a conditional probability is not operationally verifiable. Conditional probabilities, in the general case, are defined as integrands with certain properties.

This extension of the likelihood principle is supported by the Neyman-Pearson theorem, which says that, for testing the simple hypothesis H_0 that a probability distribution has a density of f_0 against the alternative hypothesis H_1 that the density function is f_1, the most powerful test of given size (probability of rejecting H_0 when true) is based on the likelihood ratio $f_1(x)/f_0(x)$. There is a critical value c; sample points which reject H_0 have $f_1(x)/f_0(x) \geq c$, and those which do not reject H_0 have $f_1(x)/f_0(x) \leq c$.

Suppose that a statistician has to make n tests of this kind, each of a simple hypothesis H_0 against a simple alternative H_1, the experiment and the hypotheses in one test having no connection with those in another. Suppose also that for the average size of his tests (probability of rejecting H_0 when true), he wishes to maximize the average power (probability of rejecting H_0 when H_1 is true). It can be shown (Pitman, 1965) that to do this, he must use the same critical value c for the likelihood ratio in all the tests. This result suggests that, to some extent, a particular numerical value of a likelihood ratio means the same thing, whatever the experiment. This is obvious when we have prior probabilities for H_0 and H_1: for if the prior odds for H_1 against

H_0 are p_1/p_0, the posterior odds are $(p_1/p_0)(f_1/f_0)$, and all that comes from the experiment is the ratio f_1/f_0. This is the argument with which we started for the discrete sample space. Its extension to the non-discrete sample space needs some sort of justification as given above, because conditional probabilities are not then directly verifiable experimentally.

Instead of the likelihood ratio $f(x, \theta_1)/f(x, \theta_0)$ we may use its logarithm, $\log f(x, \theta_1) - \log f(x, \theta_0)$, which we may call the discrimination of θ_1 against θ_0 for the point x. If θ is a real parameter, taking values in a real interval, the average rate of discrimination between θ_0 and θ_1 is

$$\frac{\log f(x, \theta_1) - \log f(x, \theta_0)}{\theta_1 - \theta_0}.$$

The limit of this when $\theta_1 \to \theta_0$, if it exists, is

$$\left.\frac{d \log f(x, \theta)}{d\theta}\right|_{\theta = \theta_0} = f'(x, \theta_0)/f(x, \theta_0),$$

where the prime denotes differentiation with respect to θ. This is the *discrimination rate* at θ_0 for the point x. It plays a central role in the theory of inference.

Suppose that for values of θ in some set A, the density function factorizes as follows.

$$f(x, \theta) = g[T(x), \theta]h(x), \qquad [1.1]$$

where T, h are functions of x only, and g, h are non-negative. For any $\theta_1, \theta_2 \in A$, the likelihood ratio

$$f(x, \theta_2)/f(x, \theta_1) = g[T(x), \theta_2]/g[T(x), \theta_1],$$

is a function of $T(x)$, and so its value is determined by the value of T. This is true for every pair of values of θ in A. Hence all the information about θ provided by the experiment is the value of $T(x)$. T is a sufficient statistic for the set A. Note that T may be real or vector-valued.

We may look at this in another way. Relative to the measure μ, $f(x, \theta)$ is the density at x of the probability measure P_θ on $\mathscr{X}$. If when [1.1] is true we use, instead of μ, the measure ν defined by $d\nu = h(x)d\mu$, the density of the probability measure becomes $g[T(x), \theta]$ at x. Clearly all the information about θ provided by the experiment is the value of $T(x)$.

Conversely, if the value of some statistic T is all the information about θ provided by the observation x, for $\theta \in A$, then for any fixed $\theta_0 \in A$, the likelihood ratio $f(x,\theta)/f(x,\theta_0)$ must be a function of T and θ only for $\theta \in A$,

$$f(x,\theta)/f(x,\theta_0) = g[T(x),\theta]$$

$$\begin{aligned} f(x,\theta) &= g[T(x),\theta]f(x,\theta_0) \\ &= g[T(x),\theta]h(x). \end{aligned}$$

If a sufficient statistic T exists for $\theta \in N$, where N is a real open interval containing θ_0,

$$L = \log f(x,\theta) = \log g[T(x),\theta] + \log h(x)$$

$$L_0' = g'[T(x),\theta_0]/g[T(x),\theta_0], \text{ where } g' = \partial g/\partial\theta.$$

The discrimination rate L_0' is a function of the sufficient statistic T.

CHAPTER TWO

DISTANCE BETWEEN PROBABILITY MEASURES

1 From the practical or experimental point of view, the problem of estimation is the problem of picking the 'actual' or 'correct' probability measure from a family of possible probability measures by observing the results of trials. It is therefore advisable first to study families of probability measures, and consider how much members of a family differ from one another.

Let P_1, P_2 be probability measures on the same σ-algebra of sets in a space $\mathscr{X}$. We want to define a measure of the difference between the two probability measures, a distance between P_1 and P_2. Suppose that they have densities f_1, f_2 relative to a dominating measure μ, which is σ-finite. This is no restriction, for we may always take $P_1 + P_2$ for μ.

Consideration of continuous distributions in R^1 suggests

$$\rho^*(P_1, P_2) = \int |f_1 - f_2| \, d\mu,$$

the L^1 norm of $f_1 - f_2$ for the distance between the distributions. This is zero when the distributions coincide and has the maximum value 2 when they do not overlap; ρ^* is the total variation of $P_1 - P_2$. For measurable sets A, $P_1(A) - P_2(A)$ is a maximum when A is the set $\{x; f_1(x) > f_2(x)\}$, and then

$$P_1(A) - P_2(A) = P_2(A^c) - P_1(A^c) = \tfrac{1}{2}\rho^*.$$

It has two main disadvantages. The modulus of a function is analytically awkward, and the L^1 norm gives the same weight to the same difference between f_1 and f_2 whether the smaller of the two is large or small. The L^2 norm of $\sqrt{f_1} - \sqrt{f_2}$ is much better in both respects, and so we define $\rho(P_1, P_2) = \rho(f_1, f_2)$, the distance between P_1 and P_2, by

$$\rho^2(P_1, P_2) = \int (\sqrt{f_1} - \sqrt{f_2})^2 d\mu = 2 - 2\int \sqrt{(f_1 f_2)} \, d\mu. \quad [2.1]$$

ρ can take values from 0 to $\sqrt{2}$. It is 0 if and only if $P_1 = P_2$. It has the maximum value $\sqrt{2}$ if and only if $f_1(x) f_2(x) = 0$ a.e. μ, i.e. if there are disjoint sets A_1, A_2 such that $P_1(A_1) = 1 = P_2(A_2)$. The distance between any discrete probability measure and any continuous probability measure has the maximum value $\sqrt{2}$, however close these measures may be in another metric.

$$\rho^* = \int|\sqrt{f_1} - \sqrt{f_2}|(\sqrt{f_1} + f_2)d\mu \geq \rho^2.$$

By Schwarz's inequality

$$\rho^{*2} \leq \int(\sqrt{f_1} - \sqrt{f_2})^2 d\mu \cdot \int(\sqrt{f_1} + \sqrt{f_2})^2 d\mu = \rho^2(4 - \rho^2).$$

Therefore

$$\rho^2 \leq \rho^* \leq \rho\sqrt{(4 - \rho^2)} \leq 2\rho.$$

The value of ρ is independent of the particular choice of the dominating measure μ. Let g_1, g_2 be the densities of P_1, P_2 relative to another dominating measure ν. Let h, k be the densities of μ, ν relative to $\mu + \nu$. The density of P_1 relative to $\mu + \nu$ is $f_1 h$ and also $g_1 k$. Thus $f_1 h = g_1 k$. Similarly $f_2 h = g_2 k$. Hence $\sqrt{(f_1 f_2)}h = \sqrt{(g_1 g_2)}k$, and

$$\begin{aligned}\int\sqrt{(g_1 g_2)}d\nu &= \int\sqrt{(g_1 g_2)}k d(\mu + \nu)\\ &= \int\sqrt{(f_1 f_2)}h d(\mu + \nu) = \int\sqrt{(f_1 f_2)}d\mu.\end{aligned}$$

Thus the distance is the same whether the densities relative to μ or to ν are used.

For any μ measurable set A,

$$\Big[\int_A \sqrt{(f_1 f_2)}d\mu\Big]^2 \leq \int_A f_1 d\mu \int_A f_2 d\mu,$$

$$\int_A \sqrt{(f_1 f_2)}d\mu \leq \sqrt{[P_1(A)P_2(A)]}.$$

Hence, if $(A_r; r = 1, \ldots, n)$ is a partition of $\mathscr{X}$ into a finite number of disjoint, measurable sets,

$$\mathscr{X} = \bigcup_{r=1}^{n} A_r,$$

$$\int\sqrt{(f_1 f_2)}d\mu \leq \sum_{r=1}^{n}\sqrt{[P_1(A_r)P_2(A_r)]}.$$

It can be shown that the left hand side of this inequality is the infimum of the right hand side for all such partitions. Thus

$$\rho^2(P_1, P_2) = 2 - 2\int \sqrt{(f_1 f_2)}d\mu = 2 - 2 \inf \sum_{r=1}^{n} \sqrt{[P_1(A_r)P_2(A_r)]}$$

$$= 2 \sup\left\{1 - \sum_{r=1}^{n} \sqrt{[P_1(A_r)P_2(A_r)]}\right\}$$

2 If T is a mapping of $\mathscr{X}$ into a space $\mathscr{Y}$, and if Q_1, Q_2 are the probability measures in $\mathscr{Y}$ induced by P_1, P_2, then

$$\rho(Q_1, Q_2) \leq \rho(P_1, P_2), \qquad [2.2]$$

with equality if and only if f_1/f_2 is a function of T, i.e. if T is a sufficient statistic.

Let g_1, g_2 be the densities of Q_1, Q_2 relative to a σ-finite measure v on $\mathscr{Y}$ which dominates the measure induced by μ. We prove [2.2] by showing that

$$\int \sqrt{(f_1 f_2)}d\mu \leq \int \sqrt{(g_1 g_2)}dv.$$

By the Schwarz inequality (iv) in Section 2 of the Appendix.

$$T^*\sqrt{(f_1 f_2)} \leq \sqrt{(T^*f_1 \cdot T^*f_2)} = \sqrt{(g_1 g_2)} \qquad \text{a.e.}v, \quad [2.3]$$

$$\int \sqrt{(f_1 f_2)}d\mu = \int T^*\sqrt{(f_1 f_2)}dv \leq \int \sqrt{(g_1 g_2)}dv,$$

with equality if and only if the equality in [2.3] holds a.e.v, i.e. if $\sqrt{f_1}/\sqrt{f_2}$ is a function of T, and therefore f_2/f_1 a function of T, so that T is a sufficient statistic. Note that a 1–1 mapping is a sufficient statistic. □ □ □

3 If

$$\Phi_{1n}(x_1, \ldots, x_n) = f_1(x_1)f_1(x_2)\ldots f_1(x_n),$$

$$\Phi_{2n}(x_1, \ldots, x_n) = f_2(x_1)f_2(x_2)\ldots f_2(x_n),$$

$$1 - \tfrac{1}{2}\rho^2(\Phi_{1n}, \Phi_{2n}) = \int \sqrt{(\Phi_{1n}\Phi_{2n})}\mu(dx_1)\ldots\mu(dx_2)$$

$$= \{\int \sqrt{(f_1 f_2)}\mu[d(x_1)]\}^n = \{1 - \tfrac{1}{2}\rho^2(f_1, f_2)\}^n,$$

and therefore $\to 0$ as $n \to \infty$, if $f_1 \neq f_2$.

Thus $\rho^2(\Phi_{1n}, \Phi_{2n}) \to 2$, the maximum value, as $n \to \infty$. This means that if two probability distributions are different, no matter how close to one another they are, by taking a sufficiently large sample, we can obtain distributions with a ρ^2, and therefore a ρ^*, as close to the maximum 2 as we please – a fundamental principle of applied statistics.

4 Some insight into the statistical significance of ρ may be gained by considering discrete distributions. Let P be the discrete probability distribution which assigns the probability p_r to the point $x_r, r = 1, 2, \ldots, k$. Let P' be the probability distribution determined by a random sample of size n from this distribution. $p'_r = n_r/n, r = 1, 2, \ldots, k$, where n_r is the number of x_r points in the sample.

Consider

$$X^2 = \sum \frac{(n_r - np_r)^2}{np_r} = n \sum \frac{(p'_r - p_r)^2}{p_r}.$$

The limit distribution of this, when $n \to \infty$, is χ^2 with $k-1$ degrees of freedom.

$$\rho^2(P, P') = \sum(\sqrt{p'_r} - \sqrt{p_r})^2 = \sum \frac{(p'_r - p_r)^2}{(\sqrt{p'_r} + \sqrt{p_r})^2}$$

$$= \frac{1}{4} \sum \frac{(p'_r - p_r)^2}{p_r} \left[1 - \frac{(\sqrt{p'_r} - \sqrt{p_r})(\sqrt{p'_r} + 3\sqrt{p_r})}{(\sqrt{p'_r} + \sqrt{p_r})^2} \right].$$

Thus

$$4n\rho^2 = n \sum \frac{(p'_r - p_r)^2}{p_r} (1 - \varepsilon_r),$$

where

$$|\varepsilon_r| \leq \frac{3|\sqrt{p'_r} - \sqrt{p_r}|}{\sqrt{p'_r} + \sqrt{p_r}},$$

and therefore $\to 0$ when $p'_r \to p_r$. The ε_r all $\to 0$ with probability one as $n \to \infty$, and

$$4n\rho^2 = X^2(1 + \eta_n),$$

where $\eta_n \to 0$ with probability one as $n \to \infty$. Hence $4n\rho^2 \to X^2$ in probability as $n \to \infty$: its limit distribution is therefore χ^2 with $k-1$ degrees of freedom.

There is no corresponding theorem for a continuous distribution. The sample distribution is discrete, and so always has maximum distance $\sqrt{2}$ from its continuous parent distribution.

Now consider two samples of sizes n_1, n_2 from the same discrete distribution P above. Let n_{1r}, n_{2r} be the numbers of

x_r points in the two samples.

$$p_{1r} = n_{1r}/n_1, \; p_{2r} = n_{2r}/n_2.$$

The usual measure of the discrepancy between the samples is

$$X^2 = n_1 n_2 \sum \frac{(n_{1r}/n_1 - n_{2r}/n_2)^2}{n_{1r} + n_{2r}} = n_1 n_2 \sum \frac{(p_{1r} - p_{2r})^2}{n_1 p_{1r} + n_2 p_{2r}}.$$

As $n_1, n_2 \to \infty$ its limit distribution is χ^2 with $k-1$ degrees of freedom.

The distance between the two sample distributions is given by

$$\rho^2 = \sum(\sqrt{p_{1r}} - \sqrt{p_{2r}})^2 = \sum \frac{(p_{1r} - p_{2r})^2}{(\sqrt{p_{1r}} + \sqrt{p_{2r}})^2}$$

$$= \frac{n_1 + n_2}{4} \sum \frac{(p_{1r} - p_{2r})^2}{n_1 p_{1r} + n_2 p_{2r}} (1 + \varepsilon_r),$$

where

$$\varepsilon_r = \frac{(\sqrt{p_{1r}} - \sqrt{p_{2r}})[(3n_1 - n_2)\sqrt{p_{1r}} - (3n_2 - n_1)\sqrt{p_{2r}}]}{(n_1 + n_2)(\sqrt{p_{1r}} + \sqrt{p_{2r}})^2}.$$

$$|\varepsilon_r| \leq \frac{3(\sqrt{p_{1r}} - \sqrt{p_{2r}})}{\sqrt{p_{1r}} + \sqrt{p_{2r}}},$$

and therefore $\to 0$ with probability one as $n_1, n_2 \to \infty$.

$$\frac{4 n_1 n_2 \rho^2}{n_1 + n_2} = X^2(1 + \eta),$$

where $\eta \to 0$ with probability one as $n_1, n_2 \to \infty$. The limit distribution of $4n_1 n_2 \rho^2/(n_1 + n_2)$ is χ^2 with $k-1$ degrees of freedom. The result in the previous case can be deduced from this case by putting $n_1 = n$, $n_2 = \infty$.

When $n_1 = n_2 = n$, the results simplify to

$$X^2 = n \sum \frac{(p_{1r} - p_{2r})^2}{p_{1r} + p_{2r}},$$

$$2n\rho^2 = n \sum \frac{(p_{1r} - p_{2r})^2}{p_{1r} + p_{2r}} \left[1 + \frac{(\sqrt{p_{1r}} - \sqrt{p_{2r}})^2}{(\sqrt{p_{1r}} + \sqrt{p_{2r}})^2}\right].$$

In this case

$$X^2 \leq 2n\rho^2 \leq 2X^2.$$

CHAPTER THREE

SENSITIVITY OF A FAMILY OF PROBABILITY MEASURES WITH RESPECT TO A PARAMETER

1 We shall consider a set $\{P_\theta;\theta\in\Theta\}$ of probability measures on the space $\mathscr{X}$ with densities $f(\cdot,\theta)$ relative to a σ-finite measure μ, where Θ is a real interval. To simplify notation we shall write f for $f(\cdot,\theta)$, and f_r for $f(\cdot,\theta_r)$ where convenient. The derivatives with respect to θ, where they exist, will be denoted by f', f_r'. For the present we denote by $\rho, = \rho(P_\theta, P_{\theta_0})$, the distance of P_θ from a fixed probability measure P_{θ_0}.

$$\rho^2 = \int(\sqrt{f}-\sqrt{f_0})^2 d\mu = 2-2\int\sqrt{(ff_0)}d\mu.$$

ρ^2, and therefore ρ, is a continuous function of θ if f is continuous in mean because $\sqrt{(ff_0)}\leq f_0+f$, and so $\sqrt{(ff_0)}$ is continuous in mean.

ρ takes the minimum value 0 at $\theta=\theta_0$. Hence if it has a derivative at θ_0, this must be zero. In general, this is not so.

$$\frac{\rho^2}{(\theta-\theta_0)^2} = \int\left(\frac{\sqrt{f}-\sqrt{f_0}}{\theta-\theta_0}\right)^2 d\mu,\ \theta\neq\theta_0. \qquad [3.1]$$

If a.e.μ, $\sqrt{f}$ has a θ derivative at θ_0, then a.e. the integrand $\to (d\sqrt{f}/d\theta)^2_{\theta=\theta_0} = f_0'^2/4f_0$ when $\theta\to\theta_0$. Under certain regularity conditions the limit of the integral will be equal to the integral of the limit and we shall have

$$\lim_{\theta\to\theta_0}\frac{\rho^2}{(\theta-\theta_0)^2} = \int\frac{f_0'^2}{4f_0}d\mu,$$

and therefore

$$\lim_{\theta\to\theta_0}\frac{\rho}{|\theta-\theta_0|} = \tfrac{1}{2}\sqrt{I_0}, \qquad [3.2]$$

where

$$I = I(\theta) = \int\frac{f'^2}{f}d\mu = E_\theta[(f'/f)^2],$$

and

$$I_0 = I(\theta_0).$$

At θ_0, ρ will have a right hand derivative $\frac{1}{2}\sqrt{I_0}$, and a left hand derivative $-\frac{1}{2}\sqrt{I_0}$. Sufficient regularity conditions are given below; but even without these, if at θ_0, f has a θ derivative a.e. then

$$\liminf_{\theta \to \theta_0} \frac{\rho^2}{(\theta - \theta_0)^2} \geq \int \frac{f_0'^2}{4f_0}\, d\mu = \tfrac{1}{4} I_0,$$

and so

$$\liminf_{\theta \to \theta_0} \frac{\rho}{|\theta - \theta_0|} \geq \tfrac{1}{2}\sqrt{I_0}. \qquad [3.3]$$

In particular, if $I_0 = \infty$, $\lim_{\theta \to \theta_0} \rho/|\theta - \theta_0| = \infty$. The right hand derivative of ρ at $\theta = \theta_0$ is $+\infty$, and the left hand derivative is $-\infty$.

A case where [3.3] is true and not [3.2] is the set of distributions on the real line with density functions

$$f(x, \theta) = e^{\theta - x}, \quad x \geq \theta,$$
$$= 0, \quad x < \theta.$$

Here

$$f'(x, \theta) = e^{\theta - x}, \quad x > \theta,$$
$$= 0, \quad x < \theta,$$

and does not exist at $x = \theta$. $I_\theta = 1$.

$$\rho^2(f, f_0) = 2 - 2e^{-|\theta - \theta_0|/2}.$$

As $\theta \to \theta_0$, $\rho^2/|\theta - \theta_0| \to 1$, and $\rho^2/(\theta - \theta_0)^2 \to \infty$, and thus $\rho/|\theta - \theta_0| \to \infty$. I_{θ_0} does not tell the full story.

2 By the mean value theorem and the theorem of dominated convergence, it follows from [3.1] that if the following conditions are satisfied, i.e.

(i) for almost all x, f is a continuous function of θ in an open interval N containing θ_0;

(ii) f has a θ derivative at all points of $N - D$, where D is a denumerable set of points in N, which does not contain θ_0, but otherwise may vary with x;

(iii) $$f'(x,\theta)^2/f(x,\theta) \leq g(x), \qquad \text{for } \theta \in N - D,$$

where g is integrable over $\mathscr{X}$;

then I_0 is finite and

$$\lim_{\theta \to \theta_0} \rho^2/(\theta - \theta_0)^2 = \tfrac{1}{4} I_0.$$

In general, this result is applicable to distributions likely to be met in practice, except when the points at which $f(x,\theta) > 0$ vary with θ. Such a case is the family of distributions with density functions

$$\begin{aligned} f(x,\theta) &= e^{\theta - x}(x-\theta)^{m-1}/\Gamma(m), \ x > \theta, && m > 0, \\ &= 0 \qquad\qquad\qquad\qquad\quad , \ x < \theta. \\ f'^2/f &= e^{\theta - x}(x-\theta)^{m-3}(x - \theta - m + 1)^2/\Gamma(m), && x > \theta, \\ &= 0, && x < \theta. \end{aligned}$$

I is finite if $m > 2$, infinite if $0 < m \leq 2$ and $m \neq 1$, and $I = 1$ if $m = 1$. The conditions set out above are satisfied when $m \geq 3$, but not when $2 < m < 3$. Many such cases can be dealt with by using the following theorem.

If $\lim \rho^2/(\theta - \theta_0)^2$ exists, is finite, and is equal to $\frac{1}{4}I_0$, we shall say that the family of probability measures (or the family of densities) is *smooth* at θ_0. If the family is smooth at every point of an open interval N, we shall say that the family is smooth in N. Instead of saying that the family of densities (or the family of probability measures) is smooth, it is sometimes convenient to say, more briefly, that f is smooth, at θ_0, or in N.

3 Theorem

If at each θ in an open interval N containing θ_0, f has a θ derivative f' at almost all x, and

(i) $\int f' d\mu = 0, = \dfrac{d}{d\theta} \int f d\mu,$

(ii) $\dfrac{d}{d\theta} \int \sqrt{(ff_0)} d\mu = \displaystyle\int \frac{d\sqrt{(ff_0)}}{d\theta} d\mu = \int \frac{f'\sqrt{f_0}}{2\sqrt{f}} d\mu,$

(iii) $I = \int f'^2/f d\mu$ *is a continuous function of* θ,
then the family is smooth at θ_0. *If also* $f'(\cdot,\theta) \xrightarrow{l} f'(\cdot,\theta_0)$*, *then at* θ_0

$$d^2\rho^2/d\theta^2 = \tfrac{1}{2}I_0,$$

where

$$\rho^2 = \rho^2(P_\theta, P_{\theta_0}).$$

Proof. Put $V = \rho^2 = \int(\sqrt{f} - \sqrt{f_0})^2 d\mu = 2 - 2\int\sqrt{(ff_0)}d\mu$.

$$V' = dV/d\theta = -\int\frac{f'\sqrt{f_0}}{\sqrt{f}}d\mu = \int\frac{(\sqrt{f}-\sqrt{f_0})f'}{\sqrt{f}}d\mu, \qquad [3.4]$$

$$0 \le \int\left(\frac{\sqrt{f}-\sqrt{f_0}}{\theta-\theta_0} - \frac{f'}{2\sqrt{f}}\right)^2 d\mu = \frac{V}{(\theta-\theta_0)^2} - \frac{V'}{\theta-\theta_0} + \tfrac{1}{4}I.$$

Hence

$$\frac{V'}{\theta-\theta_0} \le \frac{V}{(\theta-\theta_0)^2} + \tfrac{1}{4}I, \quad \theta \ne \theta_0. \qquad [3.5]$$

$V \to 0$ as $\theta \to \theta_0$. By the extended form of l'Hôpital's rule given in Section 3 of the Appendix.

$$\limsup_{\theta\to\theta_0}\frac{V}{(\theta-\theta_0)^2} \le \limsup_{\theta\to\theta_0}\frac{V'}{2(\theta-\theta_0)}.$$

From [3.5]

$$\limsup_{\theta\to\theta_0}\frac{2V}{(\theta-\theta_0)^2} \le \limsup_{\theta\to\theta_0}\frac{V}{(\theta-\theta_0)^2} + \tfrac{1}{4}I_0,$$

and therefore

$$\limsup_{\theta\to\theta_0}\frac{V}{(\theta-\theta_0)^2} \le \tfrac{1}{4}I_0.$$

From [3.3]

$$\liminf_{\theta\to\theta_0}\frac{V}{(\theta-\theta_0)^2} \ge \tfrac{1}{4}I_0,$$

*See Section 1 of the Appendix for definition of $\xrightarrow{l}$.

and so

$$\lim_{\theta \to \theta_0} \frac{V}{(\theta - \theta_0)^2} = \tfrac{1}{4} I_0.$$

This proves the first part of the theorem.

At θ_0, $V' = 0$, and so

$$\frac{V' - V'_0}{\theta - \theta_0} = \int \frac{(\sqrt{f} - \sqrt{f_0}) f'}{(\theta - \theta_0)\sqrt{f}} \, d\mu,$$

which $\to \int (f_0'^2/2f_0) d\mu = \frac{1}{2} I_0$, as $\theta \to \theta_0$, because the integrand $\overset{l}{\to} f_0'^2/2f_0$, and

$$\left| \frac{(\sqrt{f} - \sqrt{f_0}) f'}{(\theta - \theta_0)\sqrt{f}} \right| \leq \frac{(\sqrt{f} - \sqrt{f_0})^2}{(\theta - \theta_0)^2} + f'^2/f,$$

which is convergent in mean at θ_0. Thus,

at θ_0, $$dV'/d\theta = d^2 V/d\theta^2 = \tfrac{1}{2} I_0. \qquad \square\square\square$$

The distributions on R^1 with densities

$$e^{\theta - x}(x - \theta)^{m-1}/\Gamma(m), \quad m > 0, \quad x > \theta,$$

satisfy the conditions of the theorem for all $m > 2$.

4 Denote $\liminf_{\theta \to \theta_0} \rho(P_\theta, P_{\theta_0})/|\theta - \theta_0|$, which always exists, by s_0; it is a measure of the sensitivity of the family to small changes in θ at θ_0. In all cases $s_0 \geq \frac{1}{2}\sqrt{I_0}$. If s_0 is finite and equal to $\frac{1}{2}\sqrt{I_0}$, we shall say that the family is *semi-smooth* at θ_0. A smooth family is, of course semi-smooth. Semi-smoothness is a theoretical possibility rather than a contingency liable to be encountered in practice, but it must be mentioned if we are to give a complete account of I.

If the density f has a θ derivative in mean at θ_0, i.e. if f_0' exists such that

$$\lim_{\theta \to \theta_0} \int \left| \frac{f - f_0}{\theta - \theta_0} - f_0' \right| d\mu, = 0,$$

we shall say that the family of probability measures (or the family of densities) is *differentiable in mean* at θ_0.

$$\left| \frac{f - f_0}{\theta - \theta_0} \right| = \left| \frac{\sqrt{f} - \sqrt{f_0}}{\theta - \theta_0} \right| (\sqrt{f} + \sqrt{f_0}),$$

therefore

$$\left|\frac{f-f_0}{\theta-\theta_0}\right| \leq \left(\frac{\sqrt{f}-\sqrt{f_0}}{\theta-\theta_0}\right)^2 + 2f+2f_0.$$

If the family is smooth at $\theta_0, f \to f_0$ in mean, and $(\sqrt{f}-\sqrt{f_0})^2/(\theta-\theta_0)^2 \to f_0'^2/4f_0$ in mean as $\theta \to \theta_0$. Therefore $(f-f_0)/(\theta-\theta_0) \to f_0'$ in mean. Thus smoothness at θ_0 implies differentiability in mean at θ_0. For any measurable set B in $\mathscr{X}$,

$$\int_B f_0' d\mu = \frac{d}{d\theta}\int_B f d\mu |_{\theta=\theta_0}$$

which is thus a necessary condition for smoothness. In particular

$$\int f_0' d\mu = 0.$$

This last statement is true for a semi-smooth family.

5 Put

$$L = \log f,\ L_0 = \log f_0,$$
$$L' = d \log f/d\theta = f'/f,\ L_0' = f_0'/f_0.$$

When the family is smooth or semi-smooth at θ_0,

$$E_{\theta_0}(L_0') = \int (f'_0/f_0)f_0 d\mu = \int f'_0 d\mu = 0.$$
$$I_0 = E_{\theta_0}(L_0'^2) = V_{\theta_0}(L_0') = 4s_0^2.$$

If the family is smooth or semi-smooth in an open interval N, $\int f' d\mu = 0$ for all $\theta \in N$. If this can be differentiated with respect to θ,

$$\int f'' d\mu = 0,$$

$$E(\partial^2 L/\partial\theta^2) = \int \frac{\partial}{\partial\theta}(f'/f) f d\mu = (f''/f - f'^2/f^2) f d\mu,$$
$$= \int f'' d\mu - \int (f'^2/f) d\mu = -E(L'^2).$$

Thus

$$I = -E(\partial^2 L/\partial\theta^2).$$

6 Suppose we have a family of probability measures P_θ on a space $\mathscr{X}$ with densities $f(\cdot,\theta)$ relative to a σ-finite measure μ,

and a family of probability measures Q_θ on a space $\mathscr{Y}$ with densitites $g(\cdot,\theta)$ relative to a σ-finite measure ν. Consider the product measure $R_\theta = P_\theta \times Q_\theta$ on the space $\mathscr{X} \times \mathscr{Y}$. Its density $h(x,y,\theta)$ at (x,y) is $f(x,\theta)g(y,\theta)$ relative to the product measure $\mu \times \nu$.

Put

$$\rho_1 = \rho(P,P_0),\ \rho_2 = \rho(Q,Q_0),\ \rho = \rho(R,R_0)$$

then

$$1 - \tfrac{1}{2}\rho^2 = \int\sqrt{(fgf_0g_0)}\cdot d(\mu\times\nu) = \int\sqrt{(ff_0)}d\mu\cdot\int\sqrt{(gg_0)}d\nu\nu$$
$$= (1-\tfrac{1}{2}\rho_1^2)(1-\tfrac{1}{2}\rho_2^2).$$

Therefore

$$\rho^2 = \rho_1^2 + \rho_2^2 - \tfrac{1}{2}\rho_1^2\rho_2^2,$$

$$\frac{\rho^2}{(\theta-\theta_0)^2} = \frac{\rho_1^2}{(\theta-\theta_0)^2} + \frac{\rho_2^2}{(\theta-\theta_0)^2} - \frac{\rho_1^2\rho_2^2}{2(\theta-\theta_0)^2}.$$

If

$$s_{10}^2 = \lim_{\theta\to\theta_0}\frac{\rho_1^2}{(\theta-\theta_0)^2} \quad\text{and}\quad s_{20}^2 = \lim_{\theta\to\theta_0}\frac{\rho_2^2}{(\theta-\theta_0)^2},$$

exist, so does

$$s_0^2 = \lim_{\theta\to\theta_0}\frac{\rho^2}{(\theta-\theta_0)^2},$$

and

$$s_0^2 = s_{10}^2 + s_{20}^2.$$

This can be extended to any finite number of families.

$$L(x,y,\theta) = \log h(x,y,\theta) = \log f(x,\theta) + \log g(y,\theta).$$
$$L' = f'/f + g'/g.$$

If the P and Q families are both smooth at θ_0,

$$E_{\theta_0}(L_0') = E_{\theta_0}(f_0'/f_0) + E_{\theta_0}(g_0'/g_0) = 0.$$
$$\int h_0'^2/h_0\cdot d(\mu\times\nu) = V_{\theta_0}(L_0') = V_{\theta_0}(f_0'/f_0) + V_{\theta_0}(g_0'/g_0)$$
$$= 4s_{10}^2 + 4s_{20}^2 = 4s_0^2,$$

and so the R family is smooth at θ_0. This can be extended to any finite number of smooth families. In particular, if $X_1, X_2, \ldots, X_n$ are independent random variables, each with the same family of distributions, which is smooth at θ_0, then the family of $(X_1, X_2, \ldots, X_n)$ distributions is smooth at θ_0. In the same way it can be shown that if the X_r have the same semi-smooth family of distributions, the joint distribution is semi-smooth.

7 R.A. Fisher encountered I in his investigation (Fisher, 1925) of the variance of the maximum likelihood estimator in large samples. He called it the intrinsic accuracy of the distribution, and later, the amount of information in an observation. The former name has dropped out of use. I is now usually called the information; but this is not a good name, as the following examples will show.

For one observation from the normal distribution with density $e^{-(x-\theta)^2/2}/\sqrt{(2\pi)}$,

$$L = -\tfrac{1}{2}(x-\theta)^2 - \tfrac{1}{2}\log 2\pi, \ L' = x-\theta, I = 1.$$

For the densities $[1/\sqrt{(2\pi)}]e^{-(x-\theta^3)^2/2}$,

$$L = -\tfrac{1}{2}(x-\theta^3)^2 - \tfrac{1}{2}\log 2\pi, \ L' = 3\theta^2(x-\theta^3), I = 9\theta^4.$$

At $\theta=0, I=0$; but this does not mean that a single observation gives no information about θ. It merely means that $d\rho/d\theta = 0$ at $\theta = 0$, where ρ is the distance of P_θ from P_0. As a consequence of the Cramér–Rao inequality, (see Chapter 5) the θ derivative of the mean value of any statistic with a finite variance will also be 0 at $\theta = 0$.

For the exponential family $e^{\theta - x}$, $x \geq \theta$, I has the value 1, as it does for the normal family of unit variance and mean θ. For a sample of n, the value of I is n in both cases. In the normal case, the sample mean $\bar{X}_n$ is a sufficient statistic. It has mean θ and variance $1/n$. In the exponential case, the least observation $X_{(1)}$ is a sufficient statistic. $X_{(1)} - 1/n$ has mean θ and variance $1/n^2$. The two cases are quite different, although I has the same value. As shown above, I has little significance in the exponential case.

For the distributions on R^1 with densities

$$f(x,\theta) = (x-\theta)e^{\theta - x}, \qquad x \geq \theta,$$
$$= 0, \qquad x < \theta;$$

$I = \infty$ for every θ. One observation gives infinite 'information' about θ!

For the mixture of normal distributions with densities at x,

$$(1-\theta)e^{-x^2/2}/\sqrt{(2\pi)} + \theta e^{-x^2/4}/\sqrt{(4\pi)}, 0 \leq \theta \leq 1,$$

$I = \infty$ at $\theta = 0$; but for the family

$$(1-\theta^2)e^{-x^2/2}/\sqrt{(2\pi)} + \theta^2 e^{-x^2/4}/\sqrt{(4\pi)}, 0 \leq \theta \leq 1,$$

$I = 0$ at $\theta = 0$. This cannot be explained in terms of 'information'.

Evidently 'information' is a misnomer. Denoting, as before, $\liminf_{\theta \to \theta_0} \rho(P_\theta, P_{\theta_0})/|\theta - \theta_0|$, which always exists, by s_0, we shall call $4s_0^2$ the *sensitivity* of the family at θ_0. When the family is smooth or semi-smooth, or when I_0 is infinite, $I_0 = 4s_0^2$, so that then the value of I_0 is the sensitivity at θ_0. It would seem that when $I_0 < 4s_0^2$, the particular value of I_0 has no statistical significance at all. As we have seen, when independent probability distributions are combined to give a product distribution, sensitivities of smooth families are additive.

8 A mapping T of a probability space $\mathscr{X}$ into a space $\mathscr{T}$, as discussed in Chapter I, is called a statistic. The induced probability measure in $\mathscr{T}$ is called the distribution of T. For brevity we shall sometimes speak of the sensivity of a statistic T when we really mean the sensitivity of the family of distributions of T. Let the family of distributions of T be $\{Q_\theta\}$, with densities $g(\cdot, \theta)$ relative to a σ-finite measure ν. In the notation of Section 2 of the Appendix, $g = T^*f$.

Theorem

(i) *If the P_θ family is differentiable in mean at θ_0, so is the Q_θ family.*

(ii) *If the P_θ family is smooth at θ_0, or semi-smooth and differentiable in mean at θ_0, so is the Q_θ family, and the sensitivity $I_0(T)$ of T at θ_0 is less than or equal to the sensitivity I_0 of the P_θ family at θ_0, with equality if and only if $L'_0, = f'_0/f_0$, is a function of T. In Particular, if U is any statistic,*

$$I_0(L'_0, U) = I_0 = I_0(L'_0).$$

(iii) *If the P_θ family is differentiable in mean at θ_0, and L'_0 is a function of T, then*

$$g'_0(t)/g_0(t) = f'_0(x)/f_0(x), \text{ a.e., where } t = T(x).$$

Proof. If the P_θ family is differentiable in mean at θ_0, $(f-f_0)/(\theta-\theta_0) \to f'_0$ in mean as $\theta \to \theta_0$. Hence $(g-g_0)/(\theta-\theta_0) = T^*(f-f_0)/(\theta-\theta_0) \to T^*f'_0$ in mean. See Section 2 (vi) of the Appendix. Thus $T^*f'_0$ is the derivative in mean of g at θ_0. We may denote it by g'_0. For our purposes, it has all the properties of a derivative, e.g.

$$\frac{d}{d\theta}\int_B g\,dv\,|_{\theta=\theta_0} = \int_B g'_0\,dv.$$

As $\theta \to \theta_0, f \to f_0$ in mean. Therefore $g \to g_0$ in mean.

If the P_θ family is smooth at θ_0, then as $\theta \to \theta_0$,

$$\left(\frac{\sqrt{f}-\sqrt{f_0}}{\theta-\theta_0}\right)^2 \to \frac{f_0'^2}{4f_0} \text{ in mean.}$$

Therefore

$$T^*\left(\frac{\sqrt{f}-\sqrt{f_0}}{\theta-\theta_0}\right)^2 \to T^*\frac{f_0'^2}{4f_0} \quad \text{in mean.}$$

But

$$T^*\left(\frac{\sqrt{f}-\sqrt{f_0}}{\theta-\theta_0}\right)^2 = T^*\frac{f+f_0-2\sqrt{(ff_0)}}{(\theta-\theta_0)^2}$$

$$\geq \frac{g+g_0-2\sqrt{(gg_0)}}{(\theta-\theta_0)^2}, \qquad \text{a.e.}v.$$

because $T^*\sqrt{(ff_0)} \leq \sqrt{(gg_0)}$, as shown in Chapter II, Section 2. Thus

$$T^*\left(\frac{\sqrt{f}-\sqrt{f_0}}{\theta-\theta_0}\right)^2 \geq \left(\frac{\sqrt{g}-\sqrt{g_0}}{\theta-\theta_0}\right)^2 = \left(\frac{g-g_0}{\theta-\theta_0}\right)^2 \frac{1}{(\sqrt{g}+\sqrt{g_0})^2}$$

Now $(g-g_0)/(\theta-\theta_0) \to g'_0$ in mean, and $(\sqrt{g}+\sqrt{g_0})^2 \to 4g_0$ in mean.

Therefore

$$\left(\frac{\sqrt{g}-\sqrt{g_0}}{\theta-\theta_0}\right)^2 \to \frac{g_0'^2}{4g_0} \quad \text{in mean,}$$

since it converges loosely to $g_0'^2/4g_0$, and is dominated by $T^*\left(\frac{\sqrt{f}-\sqrt{f_0}}{\theta-\theta_0}\right)^2$, which is convergent in mean. The proof is the same when the P_θ family is semi-smooth and differentiable in mean at θ_0, except that attention is restricted to a sequence (θ_n) with limit θ_0, such that

$$(\sqrt{f_n}-\sqrt{f_0})^2/(\theta_n-\theta_0)^2 \to f_0'^2/4f_0 \text{ in mean.}$$

The inequality $I_0(T) \leq I_0$ is obvious from the fact that $\rho(P_\theta, P_{\theta_0}) \geq \rho(Q_\theta, Q_{\theta_0})$. To determine the conditions under which the equality holds we must use Schwarz's inequality.

$$g_0'^2 = (T^*f_0')^2 = [T^*(\sqrt{f_0}\cdot f_0'/\sqrt{f_0})]^2 \leq T^*f_0 \cdot T^*(f_0'^2/f_0) \qquad \text{a.e.}\nu.$$

Thus

$$T^*(f_0'^2/f_0) \geq g_0'^2/g_0 \qquad \text{a.e.}\nu.$$

and

$$\int f_0'^2/f_0 d\mu \geq \int g_0'^2/g_0 d\nu,$$

with equality if and only if $(f_0'/\sqrt{f_0})/\sqrt{f_0}$ is a function of T a.e.μ, i.e. $L_0' = f_0'/f_0$ is a function of T. In particular, if U is any statistic

$$I_0(L_0', U) = I_0 = I_0(L_0').$$

If T is a sufficient statistic, L_0' is a function of T, and therefore

$$I_0(T) = I_0.$$

The statement that L_0' is a function of T means only that $T(x_1) = T(x_2) \Rightarrow L_0'(x_1) = L_0'(x_2)$. If this condition is satisfied, let $L_0' = k(T)$, i.e.

$$f_0'(x)/f_0(x) = k[T(x)],\ f_0' = k(T)\cdot f_0.$$

If the P_θ family is differentiable in mean at θ_0,

$$g_0' = T^*f_0' = T^*[k(T)f_0] = k\cdot T^*f_0 = kg_0. \qquad \text{a.e.}$$

Thus $k = g_0'/g_0$, and

$$\begin{aligned} f_0'(x)/f_0(x) &= k[T(x)] = g_0'[T(x)]/g_0[T(x)] \\ &= g_0'(t)/g_0(t), \text{ if } t = T(x). \end{aligned}$$

This proves (iii), and completes the proof of the theorem.

□□□

9 The distance function ρ that we are using is based on $\sqrt{f}$; it is worth noting that smoothness of the family at θ_0 is exactly equivalent to the differentiability in mean square of $\sqrt{f}$ at θ_0. If $\sqrt{f}$ is differentiable in mean square at θ_0, g_0 (not a function of θ) exists such that

$$\lim_{\theta\to\theta_0} \int \left(\frac{\sqrt{f}-\sqrt{f_0}}{\theta-\theta_0} - g_0 \right)^2 d\mu = 0.$$

It follows that

$$\frac{\sqrt{f}-\sqrt{f_0}}{\theta-\theta_0} \xrightarrow{l} g_0 \text{ as } \theta \to \theta_0,$$

and, by the corollary to the Theorem in Section 1 of the Appendix.

$$\left(\frac{\sqrt{f}-\sqrt{f_0}}{\theta-\theta_0}\right)^2 \to g_0^2 \text{ in mean}.$$

Now

$$\frac{f-f_0}{\theta-\theta_0} = \frac{(\sqrt{f}-\sqrt{f_0})(\sqrt{f}+\sqrt{f_0})}{\theta-\theta_0} \xrightarrow{l} 2g_0\sqrt{f_0},$$

$$\left|\frac{f-f_0}{\theta-\theta_0}\right| \leq \left(\frac{\sqrt{f}-\sqrt{f_0}}{\theta-\theta_0}\right)^2 + (\sqrt{f}+\sqrt{f_0})^2 \leq \left(\frac{\sqrt{f}-\sqrt{f_0}}{\theta-\theta_0}\right)^2 + 2f + 2f_0,$$

which is convergent in mean. Hence $(f-f_0)/(\theta-\theta_0) \to 2g_0\sqrt{f_0}$ in mean: that is at θ_0, f has a derivative in mean $f_0' = 2g_0\sqrt{f_0}$, so that $g_0 = f_0'/2\sqrt{f_0}$.

$$\left(\frac{\sqrt{f}-\sqrt{f_0}}{\theta-\theta_0}\right)^2 \to f_0'^2/4f_0 \text{ in mean}:$$

the family is smooth at θ_0.

Conversely, if the family is smooth at θ_0,

$$\lim_{\theta\to\theta_0} \int \left(\frac{\sqrt{f}-\sqrt{f_0}}{\theta-\theta_0}\right)^2 d\mu = \int \frac{f_0'^2}{4f_0} d\mu,$$

where f_0' is the derivative in mean of f at θ_0.

$$\frac{\sqrt{f}-\sqrt{f_0}}{\theta-\theta_0} = \frac{f-f_0}{\theta-\theta_0} \frac{1}{\sqrt{f}+\sqrt{f_0}} \xrightarrow{l} \frac{f_0'}{2\sqrt{f_0}};$$

therefore

$$\lim_{\theta \to \theta_0} \int \left(\frac{\sqrt{f} - \sqrt{f_0}}{\theta - \theta_0} - \frac{f_0'}{2\sqrt{f_0}} \right)^2 d\mu = 0.$$

Thus $\sqrt{f}$ is differentiable in mean square at θ_0. LeCam (1970) has shown the importance of this condition in maximum likelihood theory. □□□

CHAPTER FOUR

SENSITIVITY RATING CONDITIONAL SENSITIVITY THE DISCRIMINATION RATE STATISTIC

1 We may define the relative sensitivity rating at θ_0 of two statistics as the ratio of their sensitivities at θ_0, when this has a meaning. The sensitivity rating at θ_0 of a statistic T may be defined as the ratio of its sensitivity at θ_0 to the sensitivity $4s_0^2$ of the original family of probability measures. When the family is smooth, or semi-smooth and differentiable in mean at θ_0, the sensitivity rating is $I_0(T)/I_0$.

If the family of probability measures on $\mathscr{X}$ is smooth at θ_0, and if A is an event of positive probability at θ_0, the family of conditional distributions given A is smooth at θ_0. Put $F = P_\theta(A) = \int_A f d\mu$, a function of θ only, for given A. Since the family is smooth at θ_0, it is differentiable in mean at θ_0, and therefore

$$F_0' = \int_A f_0' d\mu.$$

For the conditional distributions the space is the set A, and the density of the conditional distribution at θ is $k = f/F$, which has a derivative in mean at θ_0, $k_0' = f_0'/F_0 - f_0 F_0'/F_0^2$, because

$$\frac{f/F - f_0/F_0}{\theta - \theta_0} \xrightarrow{l} k_0' \qquad \text{as } \theta \to \theta_0,$$

and

$$\left|\frac{f/F - f_0/F_0}{\theta - \theta_0}\right| \leq f\left|\frac{1/F - 1/F_0}{\theta - \theta_0}\right| + \frac{1}{F_0}\left|\frac{f - f_0}{\theta - \theta_0}\right|,$$

which is convergent in mean. We have to show that $(\sqrt{k} - \sqrt{k_0})^2/(\theta - \theta_0)^2$, which $\xrightarrow{l} k_0'^2/4k_0$ as $\theta \to \theta_0$, is convergent in mean.

$$\frac{(\sqrt{k}-\sqrt{k_0})^2}{(\theta-\theta_0)^2} = \frac{[\sqrt{(f/F)}-\sqrt{(f_0/F_0)}]^2}{(\theta-\theta_0)^2}$$

$$\leq \frac{2[\sqrt{(f/F)}-\sqrt{(f/F_0)}]^2}{(\theta-\theta_0)^2} + \frac{2[\sqrt{(f/F_0)}-\sqrt{(f_0/F_0)}]^2}{(\theta-\theta_0)^2}$$

$$= \frac{2f(1/\sqrt{F}-1/\sqrt{F_0})^2}{(\theta-\theta_0)^2} + \frac{2(\sqrt{f}-\sqrt{f_0})^2}{F_0(\theta-\theta_0)^2}.$$

Both terms are convergent in mean in A, and so $(\sqrt{k}-\sqrt{k_0})^2/(\theta-\theta_0)^2$ is convergent in mean, and the family of conditional distributions is smooth at θ_0.

In particular, if the $\mathscr{X}$ space is discrete, all conditional distributions will be conditional on an event of positive probability, and so all conditional distributions will be smooth at θ_0.

The corresponding results can be shown to hold for a family which is semi-smooth and differentiable in mean at θ_0.

2 When the conditioning event is of zero probability, we are not able to prove so much, and we shall restrict the discussion to conditioning on the discrimination rate statistic L_0'. Let U be a statistic such that the distribution of (U, L_0') has a density relative to the product measure $\nu \times \tau$, where ν, τ are σ-finite measures in the U space and in the L_0' space (R^1) respectively. We shall show that if the family of probability measures on $\mathscr{X}$ is semi-smooth and differentiable in mean at θ_0, then for almost all c the conditional distribution of U given $L_0' = c$, has zero sensitivity at θ_0. We express this by saying that L_0' is *locally sufficient* at θ_0.

The (U, L_0') family of distributions is semi-smooth and differentiable in mean at θ_0, and has sensitivity I_0 there. Denote its density by $g(\cdot,\cdot,\theta)$. The density of the L_0' distribution is $h(\cdot,\theta)$, where

$$h(c,\theta) = \int g(\cdot, c, \theta)\, d\nu.$$

From (iii) of the Theorem in Section 8 of Chapter 3.

$$\frac{h'(c,\theta_0)}{h(c,\theta_0)} = \frac{g'(u,c,\theta_0)}{g(u,c,\theta_0)} = \frac{f'(x,\theta_0)}{f(x,\theta_0)} \quad \text{a.e.,} \qquad [4.1]$$

where $c = L'_0(x)$, $u = U(x)$. The density of the distribution of U, given $L'_0 = c$, is $k(\cdot, c, \theta)$, where

$$k(u, c, \theta) = g(u, c, \theta)/h(c, \theta). \qquad [4.2]$$

$$\liminf_{\theta \to \theta_0} \iint \frac{(\sqrt{g} - \sqrt{g_0})^2}{(\theta - \theta_0)^2}\, dv d\tau = \iint \frac{g_0'^2}{4g_0}\, dv d\tau.$$

Therefore for some sequence (θ_n) with limit θ_0

$$\iint \left| \frac{(\sqrt{g_n} - \sqrt{g_0})^2}{(\theta_n - \theta_0)^2} - \frac{g_0'^2}{4g_0} \right| dv d\tau \to 0.$$

Therefore

$$\int \left| \int \frac{(\sqrt{g_n} - \sqrt{g_0})^2}{(\theta_n - \theta_0)^2}\, dv - \int \frac{g_0'^2}{4g_0}\, dv \right| d\tau \to 0,$$

and so

$$\int \frac{(\sqrt{g_n} - \sqrt{g_0})^2}{(\theta_n - \theta_0)}\, dv \to \int \frac{g_0'^2}{4g_0}\, dv,$$

in mean on the L'_0 space.

From their convergences in mean, it follows that there exists a sequence (θ_m) with limit θ_0 such that for almost all c

$$\int \frac{(\sqrt{g_m} - \sqrt{g_0})^2}{(\theta_m - \theta_0)^2}\, dv \to \int \frac{g_0'^2}{4g_0}\, dv,$$

$$\frac{g_m - g_0}{\theta_m - \theta_0} \to g'_0 \text{ for almost all } u,$$

$$\frac{h_m - h_0}{\theta_m - \theta_0} \to h'_0.$$

$$\begin{aligned}
\frac{\sqrt{k_m} - \sqrt{k_0}}{\theta_m - \theta_0} &= \frac{\sqrt{(g_m/h_m)} - \sqrt{(g_0/h_0)}}{\theta_m - \theta_0} \\
&= \frac{1}{\sqrt{h_m}} \left(\frac{\sqrt{g_m} - \sqrt{g_0}}{\theta_m - \theta_0} - \frac{g'_0}{2\sqrt{g_0}} \right) \\
&\quad + \sqrt{g_0} \left(\frac{1/\sqrt{h_m} - 1/\sqrt{h_0}}{\theta_m - \theta_0} + \frac{h'_0}{2h_0\sqrt{h_0}} \right) \\
&\quad + \frac{1}{2} \left(\frac{g'_0}{\sqrt{(g_0 h_m)}} - \frac{h'_0\sqrt{g_0}}{h_0\sqrt{h_0}} \right).
\end{aligned}$$

Because of [4.1], the last term is equal to $(1/\sqrt{h_m} - 1/\sqrt{h_0})g_0'/2\sqrt{g_0}$, hence

$$\int \frac{(\sqrt{k_m}-\sqrt{k_0})^2}{(\theta_m-\theta_0)^2}\,dv \le \frac{3}{h_m}\int\left(\frac{\sqrt{g_m}-\sqrt{g_0}}{\theta_m-\theta_0}-\frac{g_0'}{2\sqrt{g_0}}\right)^2 dv$$
$$+\,3\left(\frac{1/\sqrt{h_m}-1/\sqrt{h_0}}{\theta_m-\theta_0}+\frac{h_0'}{2h_0^{3/2}}\right)^2\int g_0\,dv$$
$$+\,3(1/\sqrt{h_m}-1/\sqrt{h_0})^2\int\frac{g_0'^2}{4g_0}\,dv.$$

When $m \to \infty$, each of the three right-hand terms $\to 0$. Therefore the left-hand integral $\to 0$, and so

$$\liminf_{\theta\to\theta_0}\int\frac{(\sqrt{k}-\sqrt{k_0})^2}{(\theta-\theta_0)^2}\,dv = 0.$$

The conditional distribution of U given $L_0' = c$, has zero sensitivity for almost all c. □□□

3 The role of L_0' is best appreciated when the probability space is discrete. Suppose that μ is the counting measure, and that the points at which $L_0' = c$ are $x_1, x_2, \ldots$ If the P_θ family is differentiable in mean at θ_0,

$$\frac{f'(x_r,\theta_0)}{f(x_r,\theta_0)} = c = \frac{\sum f'(x_r,\theta_0)}{\sum f(x_r,\theta_0)}.$$

Put $q_s = P_\theta\{x_s \mid L_0' = c\} = f(x_s,\theta)/\sum f(x_r,\theta)$,

$$\frac{dq_s}{d\theta} = \frac{f'(x_s,\theta)}{\sum f(x_r,\theta)} - \frac{f(x_s,\theta)\sum f'(x_r,\theta)}{[\sum f(x_r,\theta)]^2}$$
$$= q_s\left\{\frac{f'(x_s,\theta)}{f(x_s,\theta)} - \frac{\sum f'(x_r,\theta)}{\sum f(x_r,\theta)}\right\}.$$

This is zero at θ_0, and thus at θ_0

$$dq_s/d\theta = 0, \quad s = 1, 2, \ldots$$

All conditional probabilities given $L_0' = c$, have zero θ derivative at θ_0. All conditional distributions will be stationary at θ_0. If the P_θ family is smooth (or semi-smooth) at θ_0, all conditional distributions, given $L_0' = c$, will be smooth (or semi-smooth), and will have zero sensitivity at θ_0.

4 If a statistic T has maximum sensitivity for all θ in an open interval N, it must be sufficient. L' must be a function of T at each θ in N, and so

$$L'(x,\theta) = k(T,\theta)$$

$$L(x,\theta) = k_1(T,\theta) + h(x), \text{ where } \partial k_1/\partial\theta = k.$$

Hence T is sufficient in N.

High sensitivity is obviously a good quality in a statistic used for estimation or testing. In the ordinary case, when the family is smooth, or semi-smooth and differentiable in mean, at θ_0, L'_0 is a statistic of maximum sensitivity at θ_0, and it is locally sufficient there. Unfortunately, in general, we cannot use L'_0 because we do not know it; it depends on the true value θ_0 of θ. It is an 'invisible' statistic. The state of affairs is different when we pass from estimation to testing. To test the hypothesis $\theta = \theta_0, L'_0$ can be used, and as is well known, the locally most powerful test is based on L'_0.

CHAPTER FIVE

EFFICACY
SENSITIVITY
THE CRAMER–RAO INEQUALITY

1 With the same notation as in Chapter III, let S be a real valued statistic which has a finite mean value $h(\theta)$ for all θ in an open interval N.

$$h(\theta) = E(S) = \int S f d\mu.$$

We shall say that S is regular at $\theta_0 \in N$ if h has a θ derivative at θ_0 given by

$$h'(\theta_0) = \int S f_0' d\mu.$$

When this is so,

$$h'(\theta_0) = \int S L_0' f_0 d\mu = E_{\theta_0}(S L_0').$$

In all applications of the theory we shall have

$$E_{\theta_0}(L_0') = \int f_0' d\mu = 0.$$

Thus when S is regular at θ_0, $h'(\theta_0)$ is the covariance of S and L_0' at θ_0, Hence

$$h'(\theta_0)^2 \leq V_{\theta_0}(L_0') V_{\theta_0}(S) = I_0 V_{\theta_0}(S),$$

where V denotes variance, and

$$I_0 = V_{\theta_0}(L_0') = E_{\theta_0}(L_0'^2) = \int f_0'^2 / f_0 d\mu.$$

Thus

$$h'(\theta_0)^2 / V_{\theta_0}(S) \leq I_0, \qquad [5.1]$$

with equality if and only if S is a linear function of L_0'. This is the Cramér–Rao inequality, which is true if the statistic S is regular. Which statistics are regular? We do not want to apply the Cramér–Rao inequality to statistics we know, and so the regularity conditions should ask as little as possible of the statistic S which we may not know, and should be mainly concerned with the family of probability measures P_θ, which we know completely.

Before discussing regularity conditions on S, consider the expression on the left of the inequality [5.1]. Its value is a measure of the effectiveness of S in the estimation of θ. We shall call it the *efficacy* of S at θ_0. The value of the numerator is an indication of the sensitivity of the distribution of S to small changes in θ at θ_0. We want this to be large. The denominator indicates the liability of S to vary from observation to observation. We want this to be small. Thus high efficacy is a desirable characteristic of a statistic used for the estimation of θ. It is unaltered by the addition of a constant to S, or by the multiplication by a non-zero constant. If L_0' itself is regular, its efficacy is I_0, the maximum possible. As will be seen later, when the family of probability measures on $\mathscr{X}$ is differentiable in mean at θ_0, the efficacy of any statistic with a finite variance at θ_0 can be defined, even if it is not regular. Even when L_0' is not regular, its efficacy is always the maximum possible.

When $h(\theta) = \theta$, so that S is an unbiased estimator of θ, the efficacy of S is simply the reciprocal of its variance,

$$V_{\theta_0}(S) \geq 1/I_0.$$

In the more general case, suppose we have a sequence (S_n) of statistics such that $E(S_n) = h(\theta)$, and $V_{\theta_0}(S_n) \to 0$ as $n \to \infty$, and that we estimate θ by taking the observed value of S_n as an estimate of $h(\theta)$, and then calculating the corresponding value of θ. Our estimator is then $\hat{\theta}_n$, where $S_n = h(\hat{\theta}_n)$.

We suppose that the S_n are all regular at θ_0, that $h(\theta)$ is strictly monotonic, and that $h'(\theta_0) \neq 0$. If θ_0 is the true value of θ, $S_n \to h(\theta_0)$ in probability as $n \to \infty$, i.e. $h(\hat{\theta}_n) \to h(\theta_0)$, and therefore $\hat{\theta}_n \to \theta_0$ in probability.

$$S_n - h(\theta_0) = h(\hat{\theta}_n) - h(\theta_0) = (\hat{\theta}_n - \theta_0)h'(\theta_0)(1 + \varepsilon_n),$$

where $\varepsilon_n \to 0$ as $\theta_n \to \theta_0$. Hence

$$\frac{h'(\theta_0)(\hat{\theta}_n - \theta_0)}{\sqrt{[V_{\theta_0}(S_n)]}} = \frac{S_n - h(\theta_0)}{\sqrt{(V_{\theta_0}(S_n))}} + \eta_n,$$

where $\eta_n \to 0$ in probability as $n \to \infty$. $[S_n - h(\theta_0)]/\sqrt{[V_{\theta_0}(S_n)]}$ has zero mean and unit variance. If it has a limit distribution, $h'(\theta_0)(\hat{\theta}_n - \theta_0)/\sqrt{[V_{\theta_0}(S_n)]}$ has the same limit distribution. If this limit distribution has zero mean and unit variance, $V_{\theta_0}(S_n)/h'(\theta_0)^2$ is the asymptotic variance of $\hat{\theta}_n$, and is the reciprocal of the efficacy of S_n.

2 Regularity conditions. Let S be a real-valued statistic with mean value $h(\theta)$. As shown above, to establish the Cramér–Rao inequality we require

(i) $h'(\theta_0) = \int S f_0' d\mu$,

(ii) $\int f_0' d\mu = 0$.

We say that S is regular at θ_0 if it satisfies (i). Equation (ii) is equivalent to the statement that the statistic with the constant value 1 is regular.

Theorem I
If the P_θ family is smooth at θ_0, every real statistic with a second moment which is bounded in some neighbourhood of θ_0 is regular at θ_0.

Proof. Let N be an open interval containing θ_0, and suppose that S has a second moment which is bounded in N, say

$$E_\theta(S^2) \le K, \quad \theta \in N.$$

Note that $S f_0'$ is integrable because $|S f_0'| \le S^2 f_0 + f_0'^2/f_0$.

$$\frac{h(\theta)-h(\theta_0)}{\theta-\theta_0} = \int \frac{S(f-f_0)}{\theta-\theta_0} d\mu,$$

$$= \int_{|S|\le c} \frac{S(f-f_0)}{\theta-\theta_0} d\mu + \int_{|S|>c} \frac{S(f-f_0)}{\theta-\theta_0} d\mu.$$

The first term on the right $\to \int_{|S|\le c} s f_0' \, d\mu$ as $\theta \to \theta_0$, because f is differentiable in mean at θ_0. Denoting the second term by $k(\theta, c)$, we have

$$k(\theta,c) = \int_{|S|>c} S(\sqrt{f}+\sqrt{f_0}) \frac{\sqrt{f}-\sqrt{f_0}}{\theta-\theta_0} d\mu.$$

Using Schwarz's inequality, and the inequality

$$(\sqrt{f}+\sqrt{f_0})^2 \le 2f + 2f_0,$$

we obtain

$$k(\theta, c)^2 \leq 4K \int_{|S|>c} \frac{(\sqrt{f}-\sqrt{f_0})^2}{(\theta-\theta_0)^2}\, d\mu,$$

which $\rightarrow K \int_{|S|>c} f_0'^2/f_0 d\mu$ as $\theta \rightarrow \theta_0$.

When $c \rightarrow \infty$, $K \int_{|S|>c} f_0'^2/f_0 d\mu \rightarrow 0$. Hence

$$\lim_{\theta\rightarrow\theta_0} \frac{h(\theta)-h(\theta_0)}{\theta-\theta_0} = \lim_{c\rightarrow\infty} \int_{|S|\leq c} Sf_0' d\mu = \int Sf_0' d\mu,$$

and S is regular at θ_0. □□□

Theorem II

Let N be a real open interval containing θ_0. If for almost all x

(i) *f is a continuous function of θ in N, and has a θ derivative f' at all points of N–D, where D is a countable set of points in N, which does not contain θ_0 but otherwise may vary with x,*

(ii) *$f'^2/f_0 \leq G$, an integrable function of x only, then every statistic S with a finite second moment at θ_0 is regular at θ_0. The conclusion still holds if* (i) *is replaced by*

(i′) *f is continuous in mean in N, and has a θ derivative in mean f' at all points of N–D, where D is a countable set which does not contain θ_0.*

Proof. Suppose that S has a finite second moment at θ_0.

$$|f'| \leq \sqrt{(f_0 G)} \qquad \text{a.e.}$$

From either (i) or (i′), the appropriate mean value theorem gives

$$\frac{|f-f_0|}{|\theta-\theta_0|} \leq \sqrt{(f_0 G)} \qquad \text{a.e.}$$

Therefore

$$\frac{|S(f-f_0)|}{|\theta-\theta_0|} \leq |S|\sqrt{(f_0 G)} \leq S^2 f_0 + G \qquad \text{a.e.}$$

The last function is integrable and independent of θ. Hence $S(f-f_0)/(\theta-\theta_0)$ converges in mean to Sf_0', and S is regular at θ_0. □□□

When we have repeated independent observations,

$x_1, x_2, \ldots, x_n, x = (x_1, x_2, \ldots, x_n)$, and

$$f(x, \theta) = \prod_1^n g(x_r, \theta),$$

where g is the density of the distribution of a single observation. It is easy to see that if g satisfies (i) or (i'), so does f. When this is so, if g satisfies (ii), then

$$\frac{|g - g_0|}{|\theta - \theta_0|} \leq \sqrt{(g_0 G)}$$

$$g \leq g_0 + \sqrt{(g_0 G)}, \text{ if } |\theta - \theta_0| \leq 1.$$

We may always take N of length ≤ 1, and then

$$g^2 \leq 2g_0^2 + 2g_0 G,$$

$$g^2/g_0 \leq 2g_0 + 2G = H,$$

integrable over the x_1 space. Also

$$g'^2/g_0 \leq G \leq H.$$

$$f'(x, \theta) = f(x, \theta) \Sigma g'(x_r, \theta)/g(x_r, \theta).$$

Hence

$$\frac{f'(x, \theta)^2}{f(x, \theta_0)} \leq n \sum \frac{f(x, \theta)^2 g'(x_r, \theta)^2}{f(x, \theta_0) g(x_r, \theta)^2}.$$

The term after the summation sign is

$$\frac{g'(x_r, \theta)^2}{g(x_r, \theta_0)} \prod_{s \neq r} \frac{g(x_s, \theta)^2}{g(x_s, \theta_0)} \leq \prod_1^n H(x_r).$$

Hence

$$f'(x, \theta)^2 / f(x, \theta_0) \leq n^2 \prod_1^n H(x_r) \qquad \text{a.e.}$$

and condition (ii) is satisfied. Thus if g satisfies condition (ii), so does f.

The normal and the Poisson densities

$$\frac{1}{\sigma\sqrt{(2\pi)}} e^{-(x-a)^2/2\sigma^2} \text{ and } \frac{e^{-\lambda}\lambda^x}{x!}$$

both satisfy the conditions of Theorem II for all their parameters.

Hence for samples from a normal or a Poisson distribution, all statistics with finite second moments are regular with respect to all parameters. For the gamma distribution density,

$$e^{(a-x)/c}(x-a)^{m-1}/c^m\Gamma(m), \qquad x \geq a; \qquad [5.2]$$

this is so for m and for c, but not for a. To deal with this we need an obvious, slight extension of Theorem II.

Theorem II′
If the condition (ii) *of Theorem II is replaced by*

(ii′) $$f'^2/f_1 \leq G,$$

where $f_1 = f(\cdot, \theta_1)$, *and* θ_1 *need not be in* N, *the conclusion is: every statistic* S *with a finite second moment at* θ_1 *is regular at* θ_0.

□□□

The gamma density [5.2] satisfies condition (ii′) if $m > 2$ and $a_1 < a$. Hence for a sample from such a gamma distribution, any statistic S is regular at a_0 if it has a finite second moment for some $a < a_0$.

3 The inequality without regularity conditions. If $E_{\theta_0}(L'_0) = 0$, and S is a real statistic, $E_{\theta_0}(SL'_0)^2 \leq V_{\theta_0}(S)V_{\theta_0}(L'_0)$ without any restrictions, with (finite) equality if and only if S is a linear function of L'_0. The only point at issue is the interpretation of $E_{\theta_0}(SL'_0)$ other than as an integral. If S is regular, and $E(S) = h(\theta)$,

$$E_{\theta_0}(SL'_0) = dE(S)/d\theta\,|_{\theta=\theta_0} = h'(\theta_0).$$

The inequality has no interest unless $V_{\theta_0}(S)$ and $V_{\theta_0}(L'_0)$ are both finite, so we assume this. $E_{\theta_0}(SL'_0)$ will then exist and be finite.

Suppose the density f is differentiable in mean at θ_0. $E_{\theta_0}(L'_0) = 0$, and any bounded statistic will be regular at θ_0. Define S_c by

$$S_c = S \text{ if } |S| \leq c,$$
$$= 0 \text{ if } |S| > c.$$

S_c has a bounded mean and variance for all θ. Let

$$E(S_c) = h_c(\theta).$$

S_c will be regular at θ_0, and

$$h'_c(\theta_0) = \int S_c f'_0 d\mu = \int_{|S| \leq c} S f'_0 d\mu.$$

Therefore

$$k(\theta_0) = \lim_{c \to \infty} h_c'(\theta_0) = \int S f_0' d\mu = E_{\theta_0}(SL_0'),$$

and

$$\frac{k(\theta_0)^2}{V_{\theta_0}(S)} \leq V_{\theta_0}(L_0') = I_0.$$

In this form, with $h'(\theta_0)$ replaced by $k(\theta_0) = \lim_{c \to \infty} h_c'(\theta_0)$, the inequality applies to every statistic which has a finite variance at θ_0, even if the statistic is not regular, provided that f is differentiable in mean at θ_0. We may then extend the definition of efficacy to every statistic S with finite variance at θ_0, by defining the efficacy of S at θ_0 as $k(\theta_0)^2/V_{\theta_0}(S)$, which is equal to $[E_{\theta_0}(SL_0')]^2/V_{\theta_0}(S)$. We shall denote the efficacy of S by $J(S)$, and its value at θ_0 by $J_0(S)$. The statistic L_0' always has efficacy equal to its variance, the maximum possible, $[E_{\theta_0}(L_0'L_0')]^2/V_{\theta_0}(L_0') = V_{\theta_0}(L_0')$. It should be noted that L_0' may not be regular. The $\Gamma(3)$ distribution with end point at θ has density $\frac{1}{2}e^{\theta - x}(x-\theta)^2$, $\theta \leq x < \infty$. $E_\theta(L_0')$ exists only for $\theta \geq \theta_0$. $E_\theta(L_0')$ has no left hand θ derivative at θ_0; but it has a right hand derivative which is equal to $E_{\theta_0}(L_0'^2)$. L_0' is semi-regular. For the symmetrical distribution with density $\frac{1}{4}e^{-|x-\theta|}(x-\theta)^2$, $-\infty < x < \infty$, $E_\theta(L_0')$ does not exist for $\theta \neq \theta_0$. L_0' is not regular.

4 The distance inequality. Let P_0, P be probability measures on $\mathscr{X}$ with densities f_0, f relative to a σ-finite measure μ. Let S be a random variable with means h_0, h and variances σ_0^2, σ^2 under P_0, P respectively.

$$h - h_0 = \int S(f - f_0) d\mu = \int (\sqrt{f} - \sqrt{f_0})(\sqrt{f} + \sqrt{f_0}) S d\mu.$$

$$(h - h_0)^2 \leq \int (\sqrt{f} - \sqrt{f_0})^2 d\mu \int S^2 (\sqrt{f} + \sqrt{f_0})^2 d\mu$$

$$\leq 2\rho^2(P, P_0) \int S^2 (f + f_0) d\mu.$$

The value of $h - h_0$ is unaltered by replacing S by $S - c$, where c is a constant. $\int (S - c)^2 (f + f_0) d\mu$ will be a minimum when $c = \frac{1}{2}(h + h_0)$. The inequality then becomes

$$(h - h_0)^2 \leq 2\rho^2[\sigma_0^2 + \sigma^2 + \tfrac{1}{2}(h - h_0)^2].$$

$$(h - h_0)^2(1 - \rho^2) \leq 2(\sigma_0^2 + \sigma^2)\rho^2.$$

When $\rho^2 < 1$, we can write this

$$(h - h_0)^2 \leq \frac{2(\sigma_0^2 + \sigma^2)\rho^2}{1 - \rho^2}, \qquad \rho < 1. \qquad [5.3]$$

This is the distance inequality relating the distance $|h - h_0|$ between the means of a random variable S, and the distance between the probability measures.

$$\frac{4\rho^2}{(h - h_0)^2} \geq \frac{1 - \rho^2}{\frac{1}{2}(\sigma_0^2 + \sigma^2)}.$$

If we have a family of measures, and $\sigma \to \sigma_0$ when $\rho \to 0$, then

$$\liminf \frac{4\rho^2}{(h - h_0)^2} \geq \frac{1}{\sigma_0^2},$$

and, if the limit exists,

$$\lim_{\rho \to 0} \frac{4\rho^2}{(h - h_0)^2} \geq \frac{1}{\sigma_0^2}.$$

Introducing a parameter θ, we have

$$\frac{2(1 - \rho^2)}{\sigma_0^2 + \sigma^2} \frac{(h - h_0)^2}{(\theta - \theta_0)^2} \leq \frac{4\rho^2}{(\theta - \theta_0)^2}.$$

If, as $\theta \to \theta_0$, $\sigma^2 \to \sigma_0^2$,

$$\frac{h'(\theta_0)^2}{\sigma_0^2} \leq 4\left(\frac{d\rho}{d\theta}\right)_{\theta_0}^2,$$

provided the limits exist. For a smooth family, this gives the Cramér–Rao inequality.

Location parameter. Let X be a random variable with finite mean and variance, and with density $f(x, \theta) = g(x - \theta)$ relative to Lebesgue measure on the real line. We may take θ to be the mean of the distribution. Let

$$\sigma^2 = \int (x - \theta)^2 g(x - \theta) dx = \int x^2 g(x) dx.$$

The inequality [5.3] becomes

$$(\theta - \theta_0)^2 \leq \frac{4\sigma^2\rho^2}{1 - \rho^2}, \quad \rho < 1.$$

$$\frac{4\rho^2}{(\theta - \theta_0)^2} \geq \frac{1 - \rho^2}{\sigma^2}.$$

Hence

$$\liminf_{\theta \to \theta_0} \frac{4\rho^2}{(\theta - \theta_0)^2} \geq \frac{1}{\sigma^2},$$

and, if the limit exists,

$$\lim_{\theta \to \theta_0} \frac{4\rho^2}{(\theta - \theta_0)^2} \geq \frac{1}{\sigma^2}.$$

The sensitivity is not less than the reciprocal of the variance. If the distribution of X is normal, the sensitivity is equal to the reciprocal of the variance. Thus, of all distributions of given variance, the normal distribution has the least sensitivity with respect to its location parameter.

The mean value of X is θ.

$$\theta = \int x g(x - \theta) dx.$$

If X is regular,

$$1 = \int x \frac{d}{d\theta} g(x - \theta) dx = \int - x g'(x - \theta) dx = - \int (x + \theta) g'(x) dx,$$

where here $g'(x)$ denotes $dg(x)/dx$. If this is true for all θ,

$$\int x g'(x) dx = -1, \quad \int g'(x) dx = 0.$$

Hence

$$1 = [\int x g'(x) dx]^2 \leq \int x^2 g(x) dx \int g'(x)^2 / g(x) dx. \qquad [5.4]$$

Thus

$$I = \int g'(x)^2 / g(x) dx \geq 1/\sigma^2,$$

the Cramér–Rao inequality in this case.

We shall have equality in [5.4] if and only if

$$g'(x)/g(x)^{1/2} = c x g(x)^{1/2}, \qquad \text{a.e., } c \text{ constant.}$$

$$g'(x)/g(x) = cx, \qquad \text{a.e.}$$

$$d \log g(x)/dx = cx, \qquad \text{a.e.}$$

$$\log g(x) = \tfrac{1}{2} c x^2 + \kappa(x),$$

where κ has a zero derivative almost everywhere.

$$g(x) = e^{cx^2/2 + \kappa(x)}.$$

We require $\int xg'(x)dx = -1$. Since $xg'(x) = cx^2g(x)$ a.e., c must be negative. Thus

$$g(x) = \frac{k(x)}{\sigma\sqrt{(2\pi)}} e^{-x^2/2\sigma^2}$$

where $k(x)$ has a derivative almost everywhere which is 0.

The function k must be non-negative, and must also be determined so that

$$\int g(x)dx = 1, \int g'(x)dx = 0, \int x^2 g(x)dx = \sigma^2.$$

X will then be regular, and $I(X) = 1/\sigma^2$.

For example, given $b > 0$, we can determine $a, c > 0$, such that if

$$\begin{aligned} k(x) &= 0, \; |x| < a, \\ &= c, \; a \leq |x| \leq a+b, \\ &= 0, \; |x| > a+b, \end{aligned}$$

the conditions will be satisfied. For simplicity we take $\sigma = 1$. The equations for a, c are

$$\frac{2c}{\sqrt{(2\pi)}} \int_a^{a+b} e^{-x^2/2} dx = 1,$$

$$\frac{2c}{\sqrt{(2\pi)}} \int_a^{a+b} x^2 e^{-x^2/2} dx = 1. \qquad [5.5]$$

Hence

$$\int_a^{a+b} x^2 e^{-x^2/2} dx = \int_a^{a+b} e^{-x^2/2} dx.$$

Integrating the left side by parts gives

$$ae^{-a^2/2} = (a+b)e^{-(a+b)^2/2}$$

For any given $b > 0$ this has a unique solution for a. Equation [5.5] then determines c.

With this g, if X has probability density $g(x-\theta)$,

$$E(X) = \theta, \; V(X) = 1, \; X \text{ is regular}, \; I = 1.$$

The sensitivity is ∞. The family of densities $g(x-\theta)$ is not smooth. The Cramér–Rao inequality is

$$V(X) \geq 1/I.$$

Here

$$V(X) = 1 = 1/I.$$

The Cramér–Rao lower bound for the variance of an unbiased, regular estimator of θ is attained.

If $X_1, X_2, \ldots$ are i.i.d.r.v. with this distribution,

$$I(X_1, X_2, \ldots, X_n) = n,$$

$\bar{X}_n = (X + \ldots + X_n)/n$ is regular, $E(\bar{X}_n) = \theta$,
$V(\bar{X}_n) = 1/n$, the Cramér–Rao lower bound.
In such a case we would not be interested in regular estimators, because there are non-regular estimators which perform much better. For example if X_{n1} is the least of a sample of n, $E(X_{n1}) = \theta + q_n$, where q_n is a function of n only. Thus $X_{n1} - q_n$ is an unbiased estimator of θ. It is not regular; but its variance is asymptotically $\alpha/n^2, n \to \infty$, α constant, and so is less than the Cramér–Rao lower bound when n is sufficiently great. It should be noted that

$$I(X_1, X_2) = 2, \; I(X_1 + X_2) = \infty.$$

Evidently, here I cannot be information.

It would seem that the Cramér–Rao inequality is of interest only when the family of probability measures is smooth. When this is so, every statistic with a variance which is bounded for θ in some neighbourhood of θ_0 is regular at θ_0. Further, it seems that $I = \int f'^2/f d\mu$ is of importance only when it is the value of the sensitivity, $\lim_{\theta \to \theta_0} 4\rho^2(P_\theta, P_{\theta_0})/(\theta - \theta_0)^2$.

5 Efficacy rating and asymptotic sensitivity rating. In order to avoid unnecessary complications, we shall suppose throughout the remainder of this chapter that the P_θ family is smooth in an open interval N, and we shall consider only values of θ in N. All induced families of distributions will be smooth in N. The statistic S discussed above will have a sensitivity $I(S)$ at θ given by $I(S) = E(g'^2/g^2)$, where g is the density of the S distribution, so that

$$J(S) \leq I(S) \leq I.$$

The efficacy of S is not greater than its sensitivity.

We define the *efficacy rating* of S at θ as $J(S)/I$, the ratio of its efficacy to the maximum possible. The statistic L'_0 has a sensitivity rating 1 at θ_0; its sensitivity at θ_0 is the same as the sensitivity at θ_0 of the original family of probability measures on $\mathscr{X}$. Its efficacy at θ_0 is the maximum possible, and so its efficacy rating at θ_0 is also 1. For any other statistic a reasonable index of its performance in estimating θ at θ_0 is the square of its correlation coefficient with L'_0 at θ_0; but this is exactly its efficacy rating at θ_0. Moreover, if S_n is a statistic which is based on n independent values of x, and which is asymptotically normal, then, under certain regularity conditions,

$$\frac{I(S_n)}{n} \to \frac{J(S_n)}{n} \qquad \text{as } n \to \infty,$$

and so

$$\frac{I(S_n)}{nI} \to \frac{J(S_n)}{nI},$$

where I is the sensitivity of the family of probability measures on $\mathscr{X}$, so that nI is the sensitivity of a sample of n. Thus the sensitivity rating of $S_n \to$ the efficacy rating of S_n. The practical value of this result comes from the fact that efficacy is often much easier to compute than sensitivity.

Consider a random variable X with a normal distribution of mean a and standard deviation c, both differentiable functions of θ. The probability density is $e^{-(x-a)^2/2c^2}/\sqrt{(2\pi)}c$.

$$L = -(x-a)^2/2c^2 - \log c - \tfrac{1}{2}\log 2\pi .$$

$$L' = (x-a)a'/c^2 + (x-a)^2c'/c^3 - c'/c .$$

The sensitivity of X is

$$I(X) = E(L'^2) = a'^2/c^2 + 2c'^2/c^2 .$$

The efficacy of X is a'^2/c^2.

If $X_1, X_2, \ldots$ are independent random variables, each with this distribution, $\bar{X}_n = \sum_r X_r/n$ is normal with mean a and standard deviation $c/n^{1/2}$. The sensitivity of $\bar{X}_n$ is therefore $na'^2/c^2 + 2c'^2/c^2$, and its efficacy is na'^2/c^2. Thus

$$\frac{I(\bar{X}_n)}{n} \to \frac{J(\bar{X}_n)}{n} \qquad \text{as } n \to \infty .$$

Consider now the general case of a sequence $X_1, X_2, \ldots$ of independent, real random variables, each with the same continuous distribution with density $f(\cdot, \theta)$ relative to Lebesgue measure. Let Y_n be a function of $X_1, \ldots, X_n$ which has mean a and standard deviation $c/\sqrt{n}$, a, c, being differentiable functions of θ. $J(Y_n) = na'^2/c^2$.

$$I(Y_n) \leq I(X_1, \ldots, X_n) = nI(X_1).$$

Hence

$$a'^2/c^2 = J(Y_n)/n \leq I(Y_n)/n \leq I(X_1).$$

Suppose that $Z_{n'} = n^{1/2}(Y_n - a)/c$, is asymptotically standard normal for all θ in N.

$$Y_n = a + cZ_n/n^{1/2}.$$

If Z_n were exactly standard normal, Y_n would be normal of mean a and standard deviation $cn^{-1/2}$, and $I(Y_n)$ would be $na'^2/c^2 + 2c'^2/c^2$. Since Z_n is approximately standard normal for all θ in N, it is reasonable to expect that $I(Y_n) = na'^2/c^2 + 2c'^2/c^2 + o(n)$. If this is so,

$$I(Y_n)/n \to a'^2/c^2 = J(Y_n)/n.$$

Let us investigate this more thoroughly.

Let $f_n(\cdot, \theta)$ be the probability density of Y_n, and $g_n(\cdot, \theta)$ that of Z_n. We assume that $f_n(y, \theta)$ is a differentiable function of y, and therefore $g_n(z, \theta)$ is a differentiable function of z.

Put $z = \sqrt{n}(y - a)/c$.

$$f_n(y, \theta) = \frac{n^{1/2}}{c} g_n(z, \theta)$$

$$f_n'(y, \theta) = \frac{n^{1/2}}{c} g_n'(z, \theta) - \frac{n^{1/2}c'}{c^2} g_n(z, \theta)$$

$$+ \frac{n^{1/2} \partial g_n(z, \theta)/\partial z}{c} \left(\frac{-n^{1/2}a'}{c} - \frac{c'z}{c} \right),$$

where, as before, a prime denotes differentiation with respect to θ.

$$\frac{I(Y_n)}{n} = \int_{-\infty}^{\infty} \frac{f_n'(y, \theta)^2}{nf_n(y, \theta)} dy = \int_{-\infty}^{\infty} \frac{c^2 f_n'(y, \theta)^2}{n^2 g_n(z, \theta)} dz.$$

Now

$$\frac{-cf_n'(y,\theta)}{n[g_n(z,\theta)]^{1/2}} = \frac{a'\partial g_n(z,\theta)/\partial z}{c[g_n(z,\theta)]^{1/2}}$$

$$+n^{-1/2}\left[\frac{c'[g_n(z,\theta)]^{1/2}}{c} + \frac{c'z\partial g_n(z)/\partial\theta}{c[g_n(z,\theta)]^{1/2}} - \frac{g_n'(z,\theta)}{[g_n(z,\theta)]^{1/2}}\right]$$

so that

$$\frac{I(Y_n)}{n} = \int_{-\infty}^{\infty}\left\{\frac{a'\partial g_n/\partial z}{cg_n^{1/2}} + \frac{1}{\sqrt{n}}\left[\frac{c'g_n^{1/2}}{c} + \frac{c'z\partial g_n/\partial z}{cg_n^{1/2}} - \frac{g_n'}{g_n^{1/2}}\right]\right\}^2 dz.$$

Using the easily proved,

$\alpha_n, \beta_n, \gamma_n, \delta_n$ all real, $\int\beta_n^2, \int\gamma_n^2, \int\delta_n^2$ all $\to 0$ as $n \to \infty$,
$\int(\alpha_n + \beta_n + \gamma_n + \delta_n)^2$ bounded $\Rightarrow \int(\alpha_n + \beta_n + \gamma_n + \delta_n)^2 \to \int\alpha_n^2$,

we obtain

$$\frac{I(Y_n)}{n} \to \frac{a'^2}{c^2}\int_{-\infty}^{\infty}\frac{(\partial g_n/\partial z)^2}{g_n}dz = \frac{J(Y_n)}{n}\int_{-\infty}^{\infty}\frac{(\partial g_n/\partial z)^2}{g_n}dz,$$

provided

(i) $c' = 0$ or $n^{-1}\displaystyle\int_{-\infty}^{\infty}\frac{z^2(\partial g_n/\partial z)^2}{g_n}dz \to 0,$

and

(ii) $n^{-1}\displaystyle\int_{-\infty}^{\infty}\frac{g_n'^2}{g_n}dz \to 0.$

Note that $I(Y_n)/n \leq I(X_1)$, and so is bounded, and that

$$\int_{-\infty}^{\infty}\left(\frac{n^{-1/2}c'g_n^{1/2}}{c}\right)^2 dz = \frac{c'^2}{nc^2}\int_{-\infty}^{\infty}g_n dz = \frac{c'^2}{nc^2} \to 0.$$

If further

(iii) $\int_{-\infty}^{\infty} \frac{(\partial g_n/\partial z)^2}{g_n} dz \to 1,$

then

$$\frac{I(Y_n)}{n} \to \frac{J(Y_n)}{n},$$

the result we are interested in. □□□

First consider the case where the distribution of Y_n depends on θ only through a location parameter. The density function is then expressible in the form

$$f_n(y,\theta) = h_n(y-a),$$

where the function h_n does not depend on θ. $c' = 0$.

$$g_n(z,\theta) = n^{-1/2}ch_n(y-a) = n^{-1/2}ch_n(n^{-1/2}cz),$$

and is therefore the same for all θ. Hence $g_n' = 0$, and conditions (i) and (ii) are satisfied.

When the distribution of Y_n depends on θ only through a location and a scale parameter, the density $f_n(y,\theta)$ is expressible in the form

$$f_n(y,\theta) = \frac{1}{c}h_n\left(\frac{y-a}{c}\right),$$

where h_n does not depend on θ. It then follows that g_n is the same for all θ, and therefore $g_n' = 0$, and (ii) is satisfied.

In all cases, Z_n has mean 0 and unit variance for all θ. Its sensitivity is $I(Z_n) = \int_{-\infty}^{\infty} g_n'^2/g_n dz$. Its limit distribution has zero sensitivity, for it is standard normal for all θ in N. We may then expect that in general $I(Z_n) \to 0$ as $n \to \infty$, or at least remains bounded. All that (ii) asks is that $I(Z_n) = o(n)$. Conditions (i) and (iii) are related, and are concerned only with the distribution of Y_n at the particular value of θ being considered. In many cases, not only is Z_n asymptotically standard normal, but its density $g_n(z,\theta)$ tends to the standard normal density $[1/\sqrt{(2\pi)}]e^{-z^2/2}$, and $\partial g_n/\partial z \to -[1/\sqrt{(2\pi)}]ze^{-z^2/2}$, so that $(\partial g_n/\partial z)^2/g_n \to [1/\sqrt{(2\pi)}]z^2e^{-z^2/2}$. Under certain regularity conditions, we shall then have

$$\int_{-\infty}^{\infty} \frac{(\partial g_n/\partial z)^2}{g_n} dz \to \frac{1}{\sqrt{(2\pi)}} \int_{-\infty}^{\infty} z^2 e^{-z^2/2} dz = 1,$$

and

$$\int_{-\infty}^{\infty} \frac{z^2(\partial g_n/\partial z)^2}{g_n} dz \to \frac{1}{\sqrt{(2\pi)}} \int_{-\infty}^{\infty} z^4 e^{-z^2/2} dz = 3.$$

The first of these is (iii), and the second is stronger than (i), which requires only

$$\int_{-\infty}^{\infty} \frac{z^2(\partial g_n/\partial z)^2}{g_n} dz = o(n).$$

6 As an example, consider the gamma distribution, with density

$$e^{-x}x^{m-1}/\Gamma(m), \ x \geq 0,$$

where $m > 0$ is a differentiable function of θ. $\bar{X}_n$ is the mean of a sample of n.

$$L = -x + (m-1)\log x - \log \Gamma(m).$$

$$L' = m' \log x - m'\psi(m), \ \psi(m) = \Gamma'(m)/\Gamma(m).$$

$$L'' = -m'^2\psi'(m) + m''(\log x - \psi(m)).$$

$$I(X) = E(-L'') = m'^2\psi'(m)$$

$$I(\bar{X}_n) = I(\Sigma X_r) = n^2 m'^2 \psi'(nm)$$

$$\frac{I(\bar{X}_n)}{n} = nm'^2\psi'(nm) = nm'^2\left(\frac{1}{nm} + \frac{2}{n^2m^2} + \dots\right)$$

$$= m'^2/m + o(1).$$

$$\frac{J(\bar{X}_n)}{n} = J(X_1) = \frac{m'^2}{m}.$$

Thus

$$\frac{I(\bar{X}_n)}{n} \to \frac{J(\bar{X}_n)}{n} = \frac{m'^2}{m}.$$

If we introduce location and scale parameters $\alpha, \gamma > 0$, which are differentiable functions of θ, and consider the distribution with density

$$\frac{e^{-(x-\alpha)/\gamma}(x-\alpha)^{m-1}}{\gamma^m\Gamma(m)}, \quad x \geq \alpha,$$

we shall still have

$$\frac{I(\bar{X}_n)}{n} \to \frac{J(\bar{X}_n)}{n} = J(X_1),$$

as before. This is because the distribution of $\sqrt{n}(\bar{X}_n - a)/c$ is unaltered: only a and c are changed. In the previous case $a = m$, $c = \sqrt{m}$. Here $a = \alpha + m\gamma$, $c = \sqrt{m\gamma}$. Thus the residual density g_n is unaltered. It must have satisfied conditions (i), (ii), (iii) before, and so must still satisfy them. Location and scale parameters can always be treated in this way whenever the statistic Y_n has the property

$$Y_n\left(\frac{x_1-\alpha}{\gamma}, \frac{x_2-\alpha}{\gamma}, \ldots, \frac{x_n-\alpha}{\gamma}\right) = \frac{Y_n(x_1, x_2, \ldots, x_n) - \alpha}{\gamma}$$

$\bar{X}_n$, $\text{Max}(X_1, X_2, \ldots, X_n)$, and $\text{Min}(X_1, X_2, \ldots, X_n)$ have this property. Note that in the example just considered

$$J(X_1) = (\alpha + m\gamma)'^2/m\gamma^2 = (\alpha' + m'\gamma + m\gamma')^2/m\gamma^2.$$

7 Median of a sample. As an example of a case where the statistic is not the sum of i.i.d. random variables, consider the median of a sample of $2n+1$. Let f be the probability density, and F the distribution function of a continuous probability distribution on the real line. Suppose that the median is at 0, $F(0) = \frac{1}{2}$, and that f is bounded, and is continuous at 0. Consider the family of distributions with densities $f(x-\theta)$ at x.

If M_{2n+1} is the median of a sample of $2n+1$, its distribution has density

$$\frac{\Gamma(2n+2)}{\Gamma(n+1)\Gamma(n+1)} f(x-\theta)F(x-\theta)^n[1-F(x-\theta)]^n.$$

$$L = \log f(x-\theta) + n\log F(x-\theta) + n\log[1-F(x-\theta)] \\ - \log B(n+1, n+1).$$

$$L' = \frac{-f'(x-\theta)}{f(x-\theta)} + \frac{n[2F(x-\theta)-1]f(x-\theta)}{F(x-\theta)[1-F(x-\theta)]}$$

where $f'(x) = \frac{d}{dx} f(x)$.

$$\frac{I(M_{2n+1})}{2n+1} = \frac{n^2\Gamma(2n+1)}{\Gamma(n+1)\Gamma(n+1)} \times \int_{-\infty}^{\infty} [2F(x)-1]^2[f(x)]^3[F(x)]^{n-2}[1-F(x)]^{n-2}dx + o(1), \quad n \to \infty.$$

Putting $y = F(x)$, $dy = f(x)dx$, we have

$$\frac{I(M_{2n+1})}{2n+1} \to \frac{4n^2\Gamma(2n+1)}{\Gamma(n+1)\Gamma(n+1)} \int_0^1 f(x)^2(y-\tfrac{1}{2})^2 y^{n-2}(1-y)^{n-2}dy.$$

If $n > 2$, $0 < \varepsilon < \frac{1}{2}$,

$$K, = \int_0^1 (y-\tfrac{1}{2})^2 y^{n-2}(1-y)^{n-2}dy, = K_1 + K_2,$$

where

$$K_1 = \int_{|y-1/2|\le\varepsilon} (y-\tfrac{1}{2})^2 y^{n-2}(1-y)^{n-2}dy, \quad K_2 = \int_{\varepsilon<|y-1/2|\le 1/2}$$

$$K_1 > \int_{|y-1/2|\le\varepsilon/2} (y-\tfrac{1}{2})^2 y^{n-2}(1-y)^{n-2}dy$$

$$> (\tfrac{1}{4}-\tfrac{1}{4}\varepsilon^2)^{n-2} \int_{|y-1/2|\le\varepsilon/2} (y-\tfrac{1}{2})^2 dy = (\tfrac{1}{4}-\tfrac{1}{4}\varepsilon^2)^{n-2}\varepsilon^3/12$$

$$K_2 < (\tfrac{1}{4}-\varepsilon^2)^{n-2} \int_0^1 (y-\tfrac{1}{2})^2 dy < (\tfrac{1}{4}-\varepsilon^2)^{n-2},$$

Therefore

$$\frac{K_2}{K_1} < \frac{12(1-4\varepsilon^2)^{n-2}}{\varepsilon^3(1-\varepsilon^2)^{n-2}} \to 0 \text{ as } n \to \infty.$$

Thus

$$K = K_1 + K_2 \sim K_1, n \to \infty.$$

Hence, when $n \to \infty$, the probability distribution P_n on $(0, 1)$ with density

$$\frac{(y-\frac{1}{2})^2 y^{n-2}(1-y)^{n-2}}{\int_0^1 (y-\frac{1}{2})^2 y^{n-2}(1-y)^{n-2}dy},$$

converges to the singular distribution with probability 1 at $\frac{1}{2}$.

When $y \to \frac{1}{2}, f(x) \to f(0)$. Hence, as $n \to \infty$,

$$\frac{\int_0^1 f(x)^2(y-\frac{1}{2})^2 y^{n-2}(1-y)^{n-2}dy}{\int_0^1 (y-\frac{1}{2})^2 y^{n-2}(1-y)^{n-2}dy} = E_{P_n}[f(X)^2] \to f(0)^2.$$

Therefore

$$\frac{I(M_{2n+1})}{2n+1} \to \frac{4n^2\Gamma(2n+1)}{\Gamma(n+1)\Gamma(n+1)} f(0)^2 \int_0^1 (y-\tfrac{1}{2})^2 y^{n-2}(1-y)^{n-2}dy$$

$$= \frac{4f(0)^2 n}{n-1} \to 4f(0)^2.$$

Thus

$$\frac{I(M_{2n+1})}{2n+1} \to 4f(0)^2.$$

$E(M_{2n+1}) = \theta + \kappa(2n+1)$, where the function κ depends on f but not on θ. Therefore $\frac{d}{d\theta}[E(M_{2n+1})] = 1$, and $J(M_{2n+1}) = 1/V(M_{2n+1})$.

$$E(M_{2n+1} - \theta)^2 = \frac{\Gamma(2n+2)}{\Gamma(n+1)\Gamma(n+1)} \int_0^1 x^2 y^n(1-y)^n dy$$

$$= \frac{\Gamma(2n+2)}{\Gamma(n+1)\Gamma(n+1)} \int_0^1 \frac{x^2}{(y-\frac{1}{2})^2}(y-\tfrac{1}{2})^2 y^n(1-y)^n dy.$$

First suppose $f(0) > 0$. $y - \frac{1}{2} = \int_0^x f(u)du$, and so

$$\lim_{y \to 1/2} \frac{y-\frac{1}{2}}{x} = f(0).$$

Hence

$$E(M_{2n+1} - \theta)^2 \sim \frac{\Gamma(2n+2)}{\Gamma(n+1)\Gamma(n+1)f(0)^2} \int_0^1 (y-\tfrac{1}{2})^2 y^n(1-y)^n dy, \quad n \to \infty,$$

$$= \frac{1}{4(2n+3)f(0)^2},$$

$(2n+1)E(M_{2n+1}-\theta)^2 \to 1/4f(0)^2$.

$$\frac{I(M_{2n+1})}{2n+1} \geq \frac{J(M_{2n+1})}{2n+1} = \frac{1}{(2n+1)V(M_{2n+1})}$$

$$\geq \frac{1}{(2n+1)E(M_{2n+1}-\theta)^2},$$

The first and the last $\to 4f(0)^2$. Therefore $J(M_{2n+1})/(2n+1) \to 4f(0)^2$. When $f(0)=0$, $I(M_{2n+1})/(2n+1) \to 0$, and therefore $J(M_{2n+1})/(2n+1) \to 0$. Thus, in all cases

$$\lim_{n\to\infty} \frac{I(M_{2n+1})}{2n+1} = 4f(0)^2 = \lim_{n\to\infty} \frac{J(M_{2n+1})}{2n+1}$$

The sensitivity rating of M_{2n+1} is

$$I(M_{2n+1})/I(X_1,\ldots,X_{2n+1}) = I(M_{2n+1})/(2n+1)I(X_1)$$
$$\to 4f(0)^2/I(X_1),$$

and the asymptotic sensitivity rating of M_{2n+1} is thus $4f(0)^2/I(X_1)$. For the Laplace family with density $\frac{1}{2}e^{-|x-\theta|}$, $-\infty < x < \infty$, $I(X_1)=1$, $f(0)=\frac{1}{2}$, and so $4f(0)^2/I(X_1)=1$; the asymptotic sensitivity rating is 1. For the Cauchy family $1/\pi[1+(x-\theta)^2]$, $4f(0)^2/I(X_1)=8/\pi^2$, and for the normal family $[1/\sigma\sqrt{(2\pi)}] \times e^{-(x-\theta)^2/2\sigma^2}$ (σ constant), the value is $2/\pi$.

8 Sensitivity and efficacy are both indicators of the rate of change of a probability measure. The former applies to distributions on any space; but the latter applies only to distributions on the real line with finite variances. To compute the sensitivity of a statistic, we need to know the density function, but computation of the efficacy requires only knowledge of the mean and variance. Also, the practical significance of efficacy is easier to grasp. For a normal distribution of fixed variance and varying mean, efficacy and sensitivity are equal. This suggests that a statistic of a sample of n, which is asymptotically normal, will have an asymptotic sensitivity rating equal to its asymptotic efficacy rating. Several examples of this are given above, but simple, sufficient conditions for this have yet to be discovered.

The efficacy of a random variable X is always less than or equal to its sensitivity. The two are equal if and only if the correlation coefficient of X with its discrimination rate L' is ± 1, i.e. if and only if L' is a linear function of X. This is true for all θ in N if and only if, in N,

$$L'(x,\theta) = a(\theta)x + b(\theta),$$

and therefore

$$L(x,\theta) = a_1(\theta)x + b_1(\theta) + h(x), \qquad [5.6]$$

where $da_1/d\theta = a$, $db_1/d\theta = b$.
The density function of X will then be

$$\exp\{a_1(\theta)x + b_1(\theta) + h(x)\}.$$

If $X_1,\ldots,X_n$ are i.i.d. random variables with this distribution, $S_n = \sum_1^n X_r$ is a sufficient statistic for θ, and $J(S_n) = I(S_n)$.

Condition [5.6] is satisfied by the normal, binomial, Poisson, and negative binomial families, shown below with their respective densities:

$$\frac{1}{\sigma\sqrt{(2\pi)}}e^{-(x-\theta)^2/2\sigma^2}, \ \sigma \text{ constant};$$

$$\binom{m}{r}p^r(1-p)^{m-r}, \ p = p(\theta);$$

$$e^{-\lambda}\lambda^r/r!, \ \lambda = \lambda(\theta);$$

$$\binom{m+r-1}{r}p^r(1-p)^m, \ p = p(\theta), \ m \text{ constant}.$$

In the case of the negative binomial family, if m also is a function of θ, the efficacy is less than the sensitivity; but it can be shown that $I(S_n)/n \to J(S_n)/n$ as $n \to \infty$.

CHAPTER SIX

MANY PARAMETERS
THE SENSITIVITY MATRIX

1 We now consider a family of probability measures $\{P_\theta\}$ on $\mathscr{X}$ with densities $f(\cdot,\boldsymbol{\theta})$ relative to a σ-finite measure μ, where $\boldsymbol{\theta}$ is a point in R^k; $\boldsymbol{\theta}'=(\theta_1,\theta_2,\ldots,\theta_k)$. We shall write $\boldsymbol{\theta}_0'=(\theta_{10},\theta_{20},\ldots,\theta_{k0})$. As before, we shall write f for $f(\cdot,\boldsymbol{\theta})$, and f_0 for $f(\cdot,\boldsymbol{\theta}_0)$ where convenient. The partial derivatives with respect to θ_r will be denoted by f_r' and f_{r0}'. $L=\log f, L_r'=\partial L/\partial\theta_r$, $L_{r0}'=L_r'(\cdot,\boldsymbol{\theta}_0)$.

Let $\boldsymbol{\theta}=\boldsymbol{\theta}_0+v\mathbf{l}$, where $\mathbf{l}'$ is a unit row vector, $\mathbf{l}'=(l_1,l_2,\ldots,l_k)$, $\Sigma l_r^2=1$, and $|v|=|\boldsymbol{\theta}-\boldsymbol{\theta}_0|$, is the distance of $\boldsymbol{\theta}$ from $\boldsymbol{\theta}_0$, $v^2=\Sigma(\theta_r-\theta_{r0})^2$. If for every fixed $\mathbf{l}$, the one parameter family $f(\cdot,\boldsymbol{\theta}_0+v\mathbf{l})$ is smooth at $v=0$, we shall say that the P_θ family (or f) is smooth at $\boldsymbol{\theta}_0$.

I_{rs} is the covariance at $\boldsymbol{\theta}$ of L_r', L_s' and $\mathbf{I}=\mathbf{I}(\boldsymbol{\theta})=(I_{rs})$ is the dispersion matrix of $L_1',L_2',\ldots,L_k'$.

Theorem

Let N be an open interval in R^k, and $\mathbf{l}'=(l_1,l_2,\ldots,l_k)$, a unit row vector. If

(i) *at every point in N, f is smooth with respect to each θ_r,*
(ii) *each partial derivative f_r' is loosely continuous in N,*
(iii) *each I_{rr} is a continuous function of $\boldsymbol{\theta}$ in N,*

then f is smooth in N, and if $\boldsymbol{\theta}=\boldsymbol{\theta}_0+v\mathbf{l}$ and $\boldsymbol{\theta}_0\in N$, then the sensitivity with respect to v at θ_0 is $\mathbf{l}'\mathbf{I}(\boldsymbol{\theta}_0)\mathbf{l}$,

$$\lim_{v\to0}\int\frac{(\sqrt{f}-\sqrt{f_0})^2}{v^2}\,d\mu=\frac{1}{4}\int\frac{(\Sigma l_r f_{r0}')^2}{f_0}\,d\mu=\tfrac{1}{4}E_{\theta_0}(\Sigma l_r L_{r0}')^2$$
$$=\tfrac{1}{4}\mathbf{l}'\mathbf{I}(\theta_0)\mathbf{l}=\tfrac{1}{2}(d^2\rho^2/dv^2)_{v=0},$$

where $\rho=\rho(f,f_0)$. Also

$$(\partial^2\rho^2/\partial\theta_r\partial\theta_s)_{\theta_0}=\tfrac{1}{2}E_{\theta_0}(L_{r0}'L_{s0}').$$

Proof. Since f is smooth with respect to θ_r, $E_\theta(L_r')=0$. Hence $I_{rs}=\int f_r'f_s'/f\,d\mu$. The continuity of I_{rs} when $r\neq s$ follows from the continuity of I_{rr} and I_{ss} be the use of the Theorem in Section 1 of the Appendix.

The density f is smooth with respect to θ_r. Hence, as shown in Section 3 of Chapter 3,

$$\frac{\partial \rho^2}{\partial \theta_r} = \int \frac{\partial}{\partial \theta_r}(\sqrt{f} - \sqrt{f_0})^2 d\mu = \int \frac{(\sqrt{f} - \sqrt{f_0}) f_r'}{\sqrt{f}} d\mu.$$

This is a continuous function of $\boldsymbol{\theta}$ in N, because f and $f_r'^2/f$ are continuous in mean, and

$$|\sqrt{f} - \sqrt{f_0}|\,|f_r'/\sqrt{f}| \leq 2f + 2f_0 + f_r'^2/f,$$

which is continuous in mean. Therefore

$$d\rho^2/dv = \Sigma l_r \partial \rho^2/\partial \theta_r = \int (\sqrt{f} - \sqrt{f_0})(\Sigma l_r f_r'/\sqrt{f}) d\mu.$$

Put $V = \rho^2$; then

$$\int \left(\frac{\sqrt{f} - \sqrt{f_0}}{v} - \Sigma l_r \frac{f_r'}{2\sqrt{f}} \right)^2 d\mu = V/v^2 - V'/v + \tfrac{1}{4}\mathbf{l}'\mathbf{I}\mathbf{l}.$$

This is similar to [3.4] in the proof of Theorem in Section 3 of Chapter 3, and the argument given there, with an obvious modification for the last one, proves all the results stated above.

□ □ □

The sensitivity of the $P_{\boldsymbol{\theta}}$ family in the direction $\mathbf{l}$ at $\boldsymbol{\theta}_0$ is $\mathbf{l}'\mathbf{I}(\boldsymbol{\theta}_0)\mathbf{l}$. The matrix $\mathbf{I} = [I_{rs}]$ has been called the information matrix; but as shown in the discussion of the case $k = 1$, this is not a good name. A better name is the *sensitivity matrix*.

$$\frac{\partial \log f}{\partial \theta_r} = \frac{1}{f} \frac{\partial f}{\partial \theta_r}.$$

If the second derivatives of f with respect to the θ_r exist, a.e.μ,

$$\frac{\partial^2 \log f}{\partial \theta_r \partial \theta_s} = \frac{1}{f} \frac{\partial^2 f}{\partial \theta_r \partial \theta_s} - \frac{1}{f^2} \frac{\partial f}{\partial \theta_r} \frac{\partial f}{\partial \theta_s}$$

$$= \frac{1}{f} \frac{\partial^2 f}{\partial \theta_r \partial \theta_s} - \frac{\partial \log f}{\partial \theta_r} \frac{\partial \log f}{\partial \theta_s}$$

$$\mathrm{E}\left(\frac{\partial \log f}{\partial \theta_r} \frac{\partial \log f}{\partial \theta_s} \right) = -\mathrm{E}\left(\frac{\partial^2 \log f}{\partial \theta_r \partial \theta_s} \right) + \int \frac{\partial^2 f}{\partial \theta_r \partial \theta_s} d\mu.$$

The last term is zero if we can differentiate twice under the integral sign in $\int f d\mu = 1$; then

$$I_{rs} = -E\left(\frac{\partial^2 \log f}{\partial\theta_r \partial\theta_s}\right).$$

If $(X_1, \dots, X_n)$ is a random sample from the distribution with density $f(x, \theta_1, \dots, \theta_k)$,

$$\mathbf{I}(X_1, \dots, X_n) = n\mathbf{I}(X_1),$$

as in the case $k = 1$.

3 The Cramér-Rao inequality for many statistics and/or many parameters.

We shall assume that the conditions (i), (ii), (iii) of the above theorem are satisfied, so that f is smooth in N. Let $S_1, \dots, S_j$ be j real statistics which are regular in N with respect to each θ_r—a sufficient condition for this is that their variances are bounded in N. Let $\mathbf{M}$ be the dispersion matrix of the j statistics. Put

$$\mathbf{S}' = (S_1, \dots, S_j),\ \mathbf{D}' = (\partial \log f/\partial\theta_1, \dots, \partial \log f/\partial\theta_k),$$

$$\mathbf{K} = E(\mathbf{DS}') = (K_{rs}),\text{ where } K_{rs} = \partial E(S_s)/\partial\theta_r = E(L'_r S_s),$$

$$\mathbf{C}' = (c_1, \dots, c_j).\ \ \mathbf{I} = E(\mathbf{DD}'),\text{ the sensitivity matrix.}$$

Consider the statistic $S = \sum_1^j c_s S_s = \mathbf{C}'\mathbf{S}$. If, as before, $\boldsymbol{\theta} = \boldsymbol{\theta}_0 + v\mathbf{l}$,

$$\frac{dE(S)}{dv} = \sum_1^k l_r \frac{\partial E(S)}{\partial\theta_r} = \sum_1^k l_r \left[\sum_1^j c_s \frac{\partial E(S_s)}{\partial\theta_r}\right] = \mathbf{l}'\mathbf{KC}.$$

$$V(S) = \mathbf{C}'\mathbf{MC}.$$

The sensitivity in the direction $\mathbf{l}$ is $\mathbf{l}'\mathbf{Il}$. Writing the Cramér-Rao inequality for S in the form

$$V(S) \geq \frac{[dE(S)/dv]^2}{\mathbf{l}'\mathbf{Il}},$$

we obtain

$$\mathbf{C}'\mathbf{MC} \geq \frac{(\mathbf{l}'\mathbf{KC})^2}{\mathbf{l}'\mathbf{Il}} \text{ for all } \mathbf{l}, \mathbf{C}. \qquad [6.1]$$

If we choose $\mathbf{l}$ so that $\mathbf{l}'\mathbf{I}\mathbf{l} = \lambda\mathbf{l}'\mathbf{K}\mathbf{C}$, i.e. $\mathbf{l} = \lambda\mathbf{I}^{-1}\mathbf{K}\mathbf{C}$, the right hand side of [6.1] becomes $\mathbf{C}'\mathbf{K}'\mathbf{I}^{-1}\mathbf{K}\mathbf{C}$. Therefore

$$\mathbf{C}'\mathbf{M}\mathbf{C} \geq \mathbf{C}'\mathbf{K}'\mathbf{I}^{-1}\mathbf{K}\mathbf{C}, \text{ for all } \mathbf{C}.$$

Thus

$$\mathbf{M} - \mathbf{K}'\mathbf{I}^{-1}\mathbf{K} \text{ is non-negative.} \qquad [6.2]$$

This is the extension of the Cramér–Rao inequality to the many statistics, many parameters case. The above argument shows that this extension simply states that the rate of change in any direction of the mean value of any linear function of the S_s obeys the simple Cramér–Rao inequality. Note that $\mathbf{I}^{-1}$ is replaced by a g-inverse $\mathbf{I}^{-}$ if $\mathbf{I}$ is singular.

It is easy to show that $\mathbf{M} - \mathbf{K}'\mathbf{I}^{-1}\mathbf{K}$ is the dispersion matrix of the statistics $\mathbf{S} - \mathbf{K}'\mathbf{I}^{-1}\mathbf{D}$, and so is non-negative. This gives a shorter, but less obvious proof of [6.2]. It does not show the relation of the extended inequality to the simple inequality for one statistic and one parameter. It does, however, show what we shall require below, that

$$\mathbf{M} - \mathbf{K}'\mathbf{I}^{-1}\mathbf{K} = 0 \qquad [6.3]$$

if and only if

$$\mathbf{S} - \mathbf{K}'\mathbf{I}^{-1}\mathbf{D} = \mathbf{a}, \text{ where } \mathbf{a} = E(\mathbf{S}). \qquad [6.4]$$

If [6.4] is true, $\mathbf{S}$ is a linear function of $\mathbf{D}$,

$$\mathbf{S} = \mathbf{G}\mathbf{D} + \mathbf{a}. \qquad [6.5]$$

Conversely, if [6.5] is true,

$$\mathbf{K}' = E(\mathbf{S}\mathbf{D}') = E(\mathbf{G}\mathbf{D}\mathbf{D}' + \mathbf{a}\mathbf{D}') = \mathbf{G}\mathbf{I}.$$

Hence $\mathbf{G} = \mathbf{K}'\mathbf{I}^{-1}$, and so [6.4] is true, and therefore [6.3].

For a single statistic S_1, $\mathbf{M} = V(S_1)$,

$$\mathbf{K}' = (\partial E(S_1)/\partial\theta_1, \ldots, \partial E(S_1)/\partial\theta_k),$$

and so

$$V(S_1) \geq \mathbf{K}'\mathbf{I}^{-1}\mathbf{K}.$$

If $j = k$, and $\mathbf{K}$ and $\mathbf{I}$ are both non-singular, $\mathbf{M} - \mathbf{K}'\mathbf{I}^{-1}\mathbf{K}$

is non-negative, and so (see Section 4 of the Appendix)

$$|\mathbf{M}| \geq |\mathbf{K}'\mathbf{I}^{-1}\mathbf{K}| = |\mathbf{K}|^2|\mathbf{I}|^{-1},$$

$$\frac{|\mathbf{K}|^2}{|\mathbf{M}|} \leq |\mathbf{I}|,$$

with equality if and only if $\mathbf{M} = \mathbf{K}'\mathbf{I}^{-1}\mathbf{K}$, i.e., as shown above, if the S_s are linear functions of the L'_r.

We define the *efficacy* of the set of statistics $(S_1, \ldots, S_k)$ as $|\mathbf{K}|^2/|\mathbf{M}|$. The *efficacy rating* is defined as $|\mathbf{K}|^2/|\mathbf{M}|\,|\mathbf{I}|$, the ratio of the efficacy to the maximum possible. The set of statistics $\mathbf{D}$ has efficacy rating 1.

In order to get some idea of the meaning of efficacy in this multi-parameter case, we diagonalize the matrices $\mathbf{M}$ and $\mathbf{K}$ by suitable linear transformations on the statistics $\mathbf{S}$ and the parameters $\boldsymbol{\theta}$. Let $\mathbf{G}$ be an orthogonal matrix such that $\mathbf{GMG}'$ is diagonal. Put

$$\mathbf{S}^* = (S_1^*, \ldots, S_k^*) = \mathbf{S}'\mathbf{G}'$$

The dispersion matrix of the statistics $\mathbf{S}^*$ is $\mathbf{M}^* = \mathbf{GMG}'$. Put

$$\boldsymbol{\phi}' = (\phi_1, \ldots, \phi_k) = \boldsymbol{\theta}'\mathbf{B},$$

where $\mathbf{B}$ is determined later. The matrices corresponding to $\mathbf{D}, \mathbf{I}, \mathbf{K}$ will be denoted by $\mathbf{D}^*, \mathbf{I}^*, \mathbf{K}^*$ respectively.

$$\begin{bmatrix} \partial/\partial\theta_1 \\ \vdots \\ \partial/\partial\theta_k \end{bmatrix} = \mathbf{B} \begin{bmatrix} \partial/\partial\phi_1 \\ \vdots \\ \partial/\partial\phi_k \end{bmatrix}.$$

Hence

$$\mathbf{D} = \mathbf{BD}^*,$$

$$\mathbf{I} = E(\mathbf{DD}') = E(\mathbf{BD}^*\mathbf{D}^{*\prime}\mathbf{B}') = \mathbf{BI}^*\mathbf{B}'.$$

$$\mathbf{K} = E(\mathbf{DS}') = E(\mathbf{BD}^*\mathbf{S}^{*\prime}\mathbf{G}) = \mathbf{BK}^*\mathbf{G}.$$

$$\mathbf{K}^* = \mathbf{B}^{-1}\mathbf{KG}'.$$

If we take

$$\mathbf{B} = \frac{\mathbf{KG}'}{|\mathbf{K}|^{1/k}},$$

$$|\mathbf{B}| = 1, \mathbf{K}^* = |\mathbf{K}|^{1/k}\mathbf{1}_k, |\mathbf{K}|^* = |\mathbf{K}|,$$

$$\frac{\partial E(S_r^*)}{\partial \phi_r} = |\mathbf{K}|^{1/k},$$

$$s \neq r, \frac{\partial E(S_r^*)}{\partial \phi_s} = 0.$$

The efficacy of S_r^* with respect to ϕ_r is $|\mathbf{K}|^{2/k}/V(S_r^*)$, and

$$\frac{|\mathbf{K}|^2}{|\mathbf{M}|} = \frac{|\mathbf{K}^*|^2}{|\mathbf{M}^*|} = \prod_{r=1}^{k} (\text{efficacy of } S_r^* \text{ with respect to } \phi_r).$$

CHAPTER SEVEN

ASYMPTOTIC POWER OF A TEST ASYMPTOTIC RELATIVE EFFICIENCY

1 Asymptotic power of a consistent test. Let θ be a real parameter of a probability distribution. Suppose that a test of the hypothesis $H_0 : \theta = \theta_0$ against the hypothesis $H_1 : \theta > \theta_0$, is that H_0 is rejected if $T_n > K_n$, where T_n is a statistic, and n is the sample number. Suppose that the size of the test is exactly or approximately α, and in the latter case $\rightarrow \alpha$ as $n \rightarrow \infty$, so that

$$P_{\theta_0}(T_n > K_n) = \alpha_n \rightarrow \alpha \text{ as } n \rightarrow \infty.$$

The power function is

$$\beta(n, \theta) = P_\theta(T_n > K_n).$$

The test is said to be *consistent* if for every $\theta > \theta_0$, $\beta(n, \theta) \rightarrow 1$ as $n \rightarrow \infty$. Suppose that the test is consistent; let us investigate the behaviour of $\beta(n, \theta)$ for large values of n. Power functions are usually difficult to evaluate, and we mostly have to be content with approximations based on limit results. For fixed $\theta > \theta_0$, $\beta(n, \theta) \rightarrow 1$ as $n \rightarrow \infty$. Hence to get a limit value less than 1, we must consider a sequence (θ_n) of θ values such that $\theta_n > \theta_0$ and $\rightarrow \theta_0$ as $n \rightarrow \infty$. We shall try to determine this sequence so that as $n \rightarrow \infty, \beta(n, \theta_n)$ tends to a given limit between α and 1.

We shall assume that for some $h > 0$ and for $\theta_0 \leq \theta \leq \theta_0 + h$, $c > 0, a(\theta), w(n)$ exist, such that as $n \uparrow \infty, w(n) \downarrow 0$, and the θ_0 distribution of

$$\frac{T_n - a(\theta_0)}{cw(n)},$$

and the θ_n distribution of

$$\frac{T_n - a(\theta_n)}{cw(n)}$$

both tend to a distribution with a continuous distribution

function F. We also assume that at $\theta_0, a(\theta)$ has derivative $a'(\theta_0) > 0$.

$$P_{\theta_0}(T_n > K_n) = P_{\theta_0}\left[\frac{T_n - a(\theta_0)}{cw(n)} > k_n\right] \to 1 - F(k_n) \to 1 - F(k),$$

where $1 - F(k) = \alpha$, and

$$k_n = \frac{K_n - a(\theta_0)}{cw(n)} \to k \text{ as } n \to \infty.$$

$$\beta(n, \theta_n) = P_{\theta_n}(T_n > K_n) = P_{\theta_n}\left[\frac{T_n - a(\theta_n)}{cw(n)} > k_n - \frac{a(\theta_n) - a(\theta_0)}{cw(n)}\right]$$
$$\to 1 - F(k - \lambda_n)$$

where

$$\lambda_n = \frac{a(\theta_n) - a(\theta_0)}{cw(n)} \sim \frac{a'(\theta_0)(\theta_n - \theta_0)}{cw(n)}.$$

In order that $\beta(n, \theta_n) \to$ a limit < 1, λ_n must $\to$ a finite limit λ. If we take $\theta_n - \theta_0 \sim \lambda cw(n)/a'(\theta_0)$, λ_n will $\to \lambda$, and $\beta(n, \theta_n) \to 1 - F(k - \lambda)$. Thus

$$\lim \beta[n, \theta_0 + \lambda cw(n)/a'(\theta_0)] = 1 - F(k - \lambda).$$

Note that if $a'(\theta_0) = 0$, and $\theta_n - \theta_0 = O[w(n)], n \to \infty$, then $\lambda_n \to 0$, and $\beta(n, \theta_n) \to 1 - F(k) = \alpha$. If $a'(\theta_0) = 0$, and the second derivative $a''(\theta_0)$ exists and is > 0,

$$\lambda_n \sim \frac{a''(\theta_0)(\theta_n - \theta_0)^2}{2cw(n)}.$$

Hence if $(\theta_n - \theta_0)^2 \sim 2\lambda cw(n)/a''(\theta_0), \lambda_n \to \lambda$ and $\beta(n, \theta_n) \to 1 - F(k - \lambda)$.

In cases encountered in practical applications, $w(n)$ is a decreasing function of n which is regularly varying at ∞, i.e. for every $b > 0$,

$$\lim_{n \to \infty} w(bn)/w(n) = b^{\gamma}.$$

The constant γ is called the exponent of regular variation. Since here $w(n)$ is a decreasing function of n, γ must be negative or zero. By far the most important case is $w(n) = n^{-1/2}$. Others of some importance are $w(n) = n^{-1}$, and $w(n) = (n \log n)^{-1/2}$.

2 If we have two tests of the same hypothesis at the same level α, and for the same power with respect to the same alternative, the first test requires a sample of n_1, and the second a sample of n_2, we may define the relative efficiency of the second test with respect to the first as n_1/n_2.

We define the *asymptotic relative efficiency*, ARE, of the second with respect to the first as

$$\lim_{n_2\to\infty} n_1/n_2 \text{ when } \beta_1(n_1, \theta_{n_1}) \sim \beta_2(n_2, \theta_{n_2}), \text{ and } \theta_{n_1} - \theta_0 \sim \theta_{n_2} - \theta_0,$$

where β_1, β_2 are the power functions. Suppose that the tests, both of asymptotic size α, are based on the statistics T_n, V_n, and that as $n \to \infty$, the θ_n distributions of

$$\frac{T_n - a_1(\theta_n)}{c_1 w(n)} \quad \text{and} \quad \frac{V_n - a_2(\theta_n)}{c_2 w(n)}$$

have the same limit distribution with continuous distribution function F, where $w(n)$ is regularly varying at ∞ with exponent $-m$. Then

$$\lim_{n\to\infty} \beta_1 [n, \theta_0 + \lambda c_1 w(n)/a'(\theta_0)] = 1 - F(k - \lambda)$$

$$= \lim_{n\to\infty} \beta_2[n, \theta_0 + \lambda c_2 w(n)/a'(\theta_0)].$$

For the same limit of power with $\theta_{n1} - \theta_0 \sim \theta_{n2} - \theta_0$, the sample sizes n_1, n_2 must be related by

$$c_1 w(n_1)/a_1'(\theta_0) \sim c_2 w(n_2)/a_2'(\theta_0).$$

$$\frac{a_2'(\theta_0)/c_2}{a_1'(\theta_0)/c_1} \sim \frac{w(n_2)}{w(n_1)} \sim \left(\frac{n_2}{n_1}\right)^{-m}$$

Hence

$$\frac{n_1}{n_2} \sim \left[\frac{a_2'(\theta_0)/c_2}{a_1'(\theta_0)/c_1}\right]^{1/m},$$

which is the ARE.

In the most important case, the limit distribution is normal, $F = \Phi$, the standard normal distribution function, $\Phi(u) = \frac{1}{\sqrt{(2\pi)}} \int_{-\infty}^{u} e^{-x^2/2} dx$. $a(\theta) = E_\theta(T_n)$. The θ_n variance of T_n is asymptotically equal to c^2/n, $w(n) = n^{-1/2}$, $m = \frac{1}{2}$. The asymptotic relative efficiency is

$$\frac{a_2'(\theta_0)^2/c_2^2}{a_1'(\theta_0)^2/c_1^2}.$$

The asymptotic normality of T_n is often easily established by using the following theorem, which, in fact, enables us to deal with a larger class of alternative hypotheses.

3 Let $g(\cdot, \theta)$, $\theta \in N$ (an open interval containing θ_0) be a family of probability densities relative to a measure ν on the real line. Let Z_{nr}, $r = 1, \ldots, n$, be n independent random variables with probability densities $g(\cdot, \theta_{nr})$ relative to ν, and let $T_n = \sum_{r=1}^{n} Z_{nr}/n$.

Denote by $a(\theta)$ and $\sigma^2(\theta)$ the mean and variance of a random variable with probability density $g(\cdot, \theta)$ relative to ν. Put

$$a_0 = a(\theta_0),\ a_{nr} = a(\theta_{nr}),\ b_n = \max_r |a_{nr}|,$$

$$\bar{a}_n = \sum_{r=1}^{n} a_{nr}/n,\ \bar{\theta}_n = \sum_{r=1}^{n} \theta_{nr}/n,$$

$$\sigma_0^2 = \sigma^2(\theta_0) > 0,\ \sigma_{nr}^2 = \sigma^2(\theta_{nr}),\ \sigma_n^2 = \sum_{r=1}^{n} \sigma_{nr}^2/n$$

$$g = g(\cdot, \theta),\ g_0 = g(\cdot, \theta_0),\ g_{nr} = g(\cdot, \theta_{nr}).$$

Theorem

If

(i) $g \xrightarrow{l} g_0$ *as* $\theta \to \theta_0$,

(ii) $\int z^2 g(z, \theta) d\nu$ *is a continuous function of* θ *at* θ_0,

(iii) $\max_r |\theta_{nr} - \theta_0| \to 0$ *as* $n \to \infty$,

then

$$\frac{T_n - \bar{a}_n}{\sigma_0 n^{-1/2}} \text{ is asymptotically standard normal.} \qquad [7.1]$$

If further

(iii′) $\max_r |\theta_{nr} - \theta_0| = O(n^{-1/2})$, $n \to \infty$,

(iv) $a(\theta)$ *has a finite derivative* $a'(\theta_0)$ *at* θ_0, *then $\bar{a}_n$ in [7.1] may be replaced by $a_0 + (\bar{\theta}_n - \theta_0)a'(\theta_0)$, so that*

$$\frac{T_n - a_0}{\sigma_0 n^{-1/2}} - \frac{(\bar{\theta}_n - \theta_0)a'(\theta_0)}{\sigma_0 n^{-1/2}}$$

is asymptotically standard normal.

Proof. It follows from (i) and (ii) and the Theorem of Section 1 of the Appendix, that $g(z,\theta), zg(z,\theta), z^2 g(z,\theta)$ are continuous in mean at θ_0. Therefore $\sigma_{nr}^2 \to \sigma_0^2$ uniformly with respect to r as $n \to \infty$. Hence $\sigma_n^2 \to \sigma_0^2$. Also $b_n \to |a_0|$.

$$\frac{T_n - \bar{a}_n}{\sigma_0 n^{-1/2}} = \frac{\sigma_n}{\sigma_0} \frac{\sum_{r=1}^{n} (z_{nr} - a_{nr})}{n^{1/2}\sigma_n}.$$

Since $\sigma_n/\sigma_0 \to 1$, we have to show that $\sum_{r=1}^{n} (z_{nr} - a_{nr})/n^{1/2}\sigma_n$ is asymptotically standard normal. This will be so if the Lindeberg condition is satisfied, namely, for every $\varepsilon > 0$,

$$W_n = \frac{1}{n\sigma_n^2} \sum_{r=1}^{n} \int_{|z - a_{nr}| > \varepsilon n^{1/2}\sigma_n} (z - a_{nr})^2 g_{nr} dv \to 0 \text{ as } n \to \infty.$$

Now $|z - a_{nr}| \le |z| + b_n$. Hence

$$W_n \le \frac{1}{n\sigma_n^2} \sum_{r=1}^{n} \int_{|z| + b_n > \varepsilon n^{1/2}\sigma_n} (2z^2 + 2b_n^2) g_{nr} dv$$

$$= \frac{2}{\sigma_n^2} \int_{|z| + b_n > \varepsilon n^{1/2}\sigma_n} (z^2 + b_n^2) g_0 dv$$

$$+ \frac{2}{n\sigma_n^2} \sum_{r=1}^{n} \int_{|z| + b_n > \varepsilon n^{1/2}\sigma_n} (z^2 + b_n^2) \times (g_{nr} - g_0) dv$$

The first term $\to 0$ as $n \to \infty$. The second term

$$\le \frac{1}{\sigma_n^2} \max_r \int (z^2 + b_n^2)(g_{nr} - g_0) dv,$$

which also $\to 0$ as $n \to \infty$. This proves the first part of the theorem. This result, of course, includes the result that $(T_n - a_0)/\sigma_0 n^{-1/2}$ is asymptotically standard normal when $\theta_{nr} = \theta_0$, for all r, n.

When (iii′) and (iv) are true,

$$\bar{a}_n - a_0 = \sum_{r=1}^{n} (a_{nr} - a_0)/n = \sum_{r=1}^{n} (\theta_{nr} - \theta_0)[a'(\theta_0) + \varepsilon_{nr}]/n,$$

where $\max_r |\varepsilon_{nr}| \to 0$ as $n \to \infty$. Thus

$$\bar{a}_n - a_0 = (\bar{\theta}_n - \theta_0)a'(\theta_0) + \varepsilon_n \max_r |\theta_{nr} - \theta_0|,$$

where $\varepsilon_n \to 0$ as $n \to \infty$.

$$\frac{T_n - \bar{a}_n}{\sigma_0 n^{-1/2}} = \frac{T_n - a_0 - (\bar{\theta}_n - \theta_0)a'(\theta_0)}{\sigma_0 n^{-1/2}} + \frac{\varepsilon_n \max_r |\theta_{nr} - \theta_0| n^{1/2}}{\sigma_0}.$$

The last term $\to 0$ as $n \to \infty$.

□□□

In order to simplify the exposition, we have introduced the family $g(\cdot, \theta)$ of probability density functions of the distributions on the real line; but we may not know g. Often we start with a space $\mathscr{X}$ on which there is a family of probability measures with densities $f(\cdot, \theta)$ relative to a measure μ, and Z is a known random variable on $\mathscr{X}$. Denoting its probability density relative to the measure ν on the real line by $g(\cdot, \theta)$, we can verify conditions (i) and (ii) without knowing g. Condition (i) is equivalent to g being convergent in mean at θ_0. This will be so if f is convergent in mean there. Condition (ii) simply states that Z has a second moment which is continuous at θ_0. This can be determined from a knowledge of f and Z only, without knowing g.

4 Let g be a probability density relative to Lebesgue measure on the real line, G the corresponding distribution function. Suppose that $G(0) = 0, g(0) > 0$, and that g has right hand continuity at 0. Consider the family of probability measures with densities $g(x - \theta)$ at x.

Let T_n be the least member in a sample of n.

$$P_\theta(T_n - \theta > x) = [1 - G(x)]^n.$$

$$P_\theta[(T_n - \theta)ng(0) > x] = [1 - G(x/ng(0))]^n.$$

When $n \to \infty, x/ng(0) \to 0$, and $G[x/ng(0)] \sim [x/ng(0)]g(0) = x/n$. Therefore

$$\{1 - G[x/ng(0)]\}^n \to (1 - x/n)^n \to e^{-x}.$$

Thus

$$P_\theta[(T_n - \theta)ng(0) \leq x] = 1 - \{1 - G[x/ng(0)]\}^n \to 1 - e^{-x}, x \geq 0$$

Here $\qquad a(\theta) = \theta,\ c = 1/g(0),\ w(n) = n^{-1}.$

For testing $\theta = \theta_0$ against $\theta > \theta_0$, at level $\alpha = e^{-k}$,

$$\beta[n, \theta_0 + \lambda/ng(0)] \to e^{\lambda - k},\ \lambda < k,$$
$$\to 1, \quad \lambda \geq k.$$

CHAPTER EIGHT

MAXIMUM LIKELIHOOD ESTIMATION

1 General results. We consider a family of probability measures on $\mathscr{X}$ dominated by a σ-finite measure μ. The corresponding density functions are $f(\cdot,\theta), \theta\in\Theta$, where Θ is a set in R^k. We shall denote the likelihood function for a set of n observations by f_n, so that

$$f_n(x_1, x_2, \dots, x_n, \theta) = \prod_{r=1}^{n} f(x_r, \theta).$$

For any set A which intersects Θ, we write

$$f^*(x, A) = \sup\{f(x,\theta); \theta\in A\Theta\},$$

$$f_n^*(x_1, x_2, \dots, x_n, A) = \sup\{f_n(x_1, x_2, \dots, x_n, \theta); \theta\in A\Theta\}.$$

For the proof of the main theorem we need the following three lemmas.

Lemma 1

If f is a probability density relative to a measure μ, and if g is a density or a subdensity, i.e. $g \geq 0, \int g d\mu \leq 1$, and if $\rho(f,g) > 0$, then $E_f \log(g/f) < 0$.

Proof If $Z > 0, \log Z \leq Z - 1$, with equality if, and only if, $Z = 1$. Therefore

$$E_f \log(g/f) = \int \log(g/f) f d\mu < \int (g/f - 1) f d\mu \leq \int (g - f) d\mu = 0.$$

□□□

Lemma 2

For any set A with intersects Θ,

$$E_{\theta_0} \log \frac{f_{r+1}(\cdot, \theta_0)}{f_{r+1}^*(\cdot, A)} \geq \frac{r+1}{r} E_{\theta_0} \log \frac{f_r(\cdot, \theta_0)}{f_r^*(\cdot, A)},$$

when the right hand side exists, finite or infinite.

Proof

$$f_{r+1}(x_1, x_2, \ldots, x_{r+1}, \theta)^r = \prod f_r(x_2, x_3, \ldots, x_{r+1}, \theta),$$

where the factors in the product are the $r+1$ likelihoods of r out of $x_1, x_2, \ldots, x_{r+1}$. Hence

$$f_{r+1}^*(x_1, x_2, \ldots, x_{r+1}, A)^r \leq \prod f_r^*(x_2, x_3, \ldots, x_{r+1}, A).$$

Therefore

$$r \log \frac{f_{r+1}(x_1, x_2, \ldots, x_{r+1}, \theta_0)}{f_{r+1}^*(x_1, x_2, \ldots, x_{r+1}, A)} \geq \sum \log \frac{f_r(x_2, x_3, \ldots, x_{r+1}, \theta_0)}{f_r^*(x_2 x_3, \ldots, x_{r+1}, A)}.$$

Hence

$$rE_{\theta_0} \log \frac{f_{r+1}(\cdot, \theta_0)}{f_{r+1}^*(\cdot, A)} \geq (r+1) E_{\theta_0} \log \frac{f_r(\cdot, \theta_0)}{f_r^*(\cdot, A)},$$

if the right-hand side exists.

□ □ □

Corollary

If for $r = m$

$$E_{\theta_0} \log \frac{f_r(\cdot, \theta_0)}{f_r^*(\cdot, A)} > 0 \text{ (alternatively } > -\infty)$$

then this is true for all $r \geq m$.

Lemma 3

If for a set A which intersects Θ, and for some value of r,

$$E_{\theta_0} \log \frac{f_r(X_1, \ldots, X_r, \theta_0)}{f_r^*(X_1, \ldots, X_r, A)} > 0, \qquad [8.1]$$

then, with probability one,

$$f_n(X_1, \ldots, X_n, \theta_0) > f_n^*(X_1, \ldots, X_n, A) \qquad [8.2]$$

when n is great.

Proof. To simplify the printing let us denote

$$\log \frac{f_{s-r+1}(X_r, \ldots, X_s, \theta_0)}{f_{s-r+1}^*(X_r, \ldots, X_s, A)}$$

by $W(r, s)$, $r \leq s$.

Note that if $r \leq s < t$,

$$f_{t-r+1}(X_r, \ldots, X_t, \theta) = f_{s-r+1}(X_r, \ldots, X_s, \theta) f_{t-s}(X_{s+1}, \ldots, X_t, \theta),$$

and therefore

$$f^*_{t-r+1}(X_r, \ldots, X_t, A) \leq f^*_{s-r+1}(X_r, \ldots, X_s, A) f^*_{t-s}(X_{s+1}, \ldots, X_t, A).$$

Hence $W(r, t) \geq W(r, s) + W(s + 1, t)$.

Suppose [8.1] is true when $r = m$, then from the corollary to Lemma 2 it is true for $r \geq m$. When $n \geq 2m$, there are positive integers u, v such that

$$n = vm + u, \quad m \leq u < 2m.$$

$$\log \frac{f_n(X_1, \ldots, X_n, \theta_0)}{f_n^*(X_1, \ldots, X_n, A)} = W(1, n)$$

$$W(1, n) \geq W(1, u) + \sum_{r=0}^{v-1} W(rm + u + 1, rm + u + m).$$

Since $u \geq m$, for fixed u, the first term on the right has a mean value > 0, and therefore is $> -\infty$ with probability one. The v random variables under the summation sign are independent, and each is distributed like $W(1, m)$ with a positive mean value. It follows from the strong law of large numbers that, with probability one, their sum $\rightarrow \infty$ as $v \rightarrow \infty$. Hence, for fixed u,

$$\lim_{v \rightarrow \infty} W(1, vm + u) = \infty, \text{ a.s.}$$

This is true for each of the m values of u, and therefore

$$\lim_{n \rightarrow \infty} W(1, n) = \infty, \text{ a.s.}$$

Hence

$$f_n(X_1, \ldots, X_n, \theta_0)/f_n^*(X_1, \ldots, X_n, A) \rightarrow \infty, \text{ a.s.},$$

and so [8.2] follows. We shall say that a set A is inferior to θ_0 if it satisfies [8.2]. The union of a finite number of sets inferior to θ_0 is inferior to θ_0.

□□□

Theorem

Let $X_1, X_2, \ldots, X_n$ be i.i.d. random elements each with probability density $f(x, \theta_0)$ at x. We assume that

(i) *if* $\theta \neq \theta_0$, $\rho(f, f_0) > 0$;

(ii) *for each x the density f is an upper semi-continuous function of* θ *in* Θ, *i.e. if* $\phi \in \Theta$,

$$\lim_{h \to 0} \sup[f(x, \theta); |\theta - \phi| < h] = f(x, \phi).$$

If H is a compact subset of Θ *which contains* θ_0, *and if for some value of r*

(iii) $$E_{\theta_0} \log \frac{f_r(\cdot, \theta_0)}{f_r^*(\cdot, H)} > -\infty,$$

then, $\hat{\theta}_n \in H$ *exists, such that*

$$f_n(x_1, x_2, \ldots, x_n, \hat{\theta}_n) = f_n^*(x_1, x_2, \ldots, x_n, H),$$

and with probability one, $\hat{\theta}_n \to \theta_0$ *as* $n \to \infty$. *If in addition,*

(iv) $$E_{\theta_0} \log \frac{f_r(\cdot, \theta_0)}{f_r^*(\cdot, H^c)} > 0,$$

then with probability one, at $\hat{\theta}_n$ *the likelihood function has a global maximum, when n is great.*

$$f_n(x_1, x_2, \ldots, x_n, \hat{\theta}_n) = f_n^*(x_1, x_2, \ldots, x_n, \Theta).$$

Some sort of continuity condition of f is necessary. While this is not the main reason for imposing it, condition (ii) does result in the supremum of $f_r(x_1, \ldots, x_r, \theta)$ in any compact set $H \subset \Theta$ being attained for some $\theta \in H$. In practical cases the satisfaction of this condition can usually be achieved by suitable definition of $f(x, \theta)$ at points of discontinuity with respect to θ.

Condition (iii) rules out densities like $e^{\theta - x}(x - \theta)^{-1/2}/\Gamma(\frac{1}{2})$, which have infinities whose position varies with θ. Such cases need special treatment, and are hardly likely to be met in practice.

A sufficient condition for (iii) is

(iii′) $h(x_1, \ldots, x_r) f_r(x_1, \ldots, x_r, \theta)$ bounded for all θ in H and all $x_1, \ldots, x_r$,

and

$$E_{\theta_0} \log[h(X_1, \ldots, X_r) f_r(X_1, \ldots, X_r, \theta_0)] > -\infty,$$

where

$$h(x_1, \ldots, x_r) > 0.$$

Often $h \equiv 1$. Suppose

$$h(x_1,\dots,x_r)f_r(x_1,\dots,x_r,\theta) \le C,$$

then

$$h(x_1,\dots,x_r)f_r^*(x_1,\dots,x_r,H) \le C.$$

$$E_{\theta_0}\log\frac{f_r(\cdot,\theta_0)}{f_r^*(\cdot,H)} = E_{\theta_0}\log[h\cdot f_r(\cdot,\theta_0)] - E_{\theta_0}\log[h\cdot f_r^*(\cdot,H)]$$
$$> -\infty - \log C.$$

Sufficient conditions for (iii) and (iv) are (iii″) and (iv′).

(iii″) For some r, $\quad E_{\theta_0}\log\dfrac{f_r(\cdot,\theta_0)}{f_r^*(\cdot,\Theta)} > -\infty.$

(iv′) There exists an expanding sequence (H_m) of compact sets in Θ such that for some r and for almost all x,

$$f_r^*(x_1,\dots,x_r,H_m^c) \to \kappa(x_1,\dots,x_r), \text{ as } m \to \infty,$$

where

$$\int \kappa\, d\mu_r \le 1, \quad \text{and} \quad \rho(\kappa, f_r(\cdot,\theta_0)) > 0. \quad \text{Often} \quad \kappa \equiv 0.$$

The condition (iii″) implies (iii), and when (iii″) is true,

$$-\infty < E_{\theta_0}\log\frac{f_r(\cdot,\theta_0)}{f_r^*(\cdot,H_m^c)}.$$

$$\log\{f_r(\cdot,\theta_0)/f_r^*(\cdot,H_m^c)\} \uparrow \log[f_r(\cdot,\theta_0)/\kappa] \text{ when } m\uparrow\infty.$$

Hence

$$E_{\theta_0}\log\frac{f_r(\cdot,\theta_0)}{f_r^*(\cdot,H_m^c)} \uparrow E_{\theta_0}\log\frac{f_r(\cdot,\theta_0)}{\kappa},$$

which is >0 by Lemma 1. Thus (iv) will be true for H_m when m is sufficiently great. Clearly θ_0 will be an element of such an H_m.

Proof. Let H be a compact subset of Θ which contains θ_0 and at least one other point of Θ. For any $h>0$, let $N_h(\phi)$ denote the open ball in R^k with the centre ϕ and radius h. Take $h_0>0$ and sufficiently small so that $K = H - N_{h_0}(\theta_0)$ is not empty. K is compact.

Suppose (iii) is true when $r=m$. If $\phi \in K$,

$$-\infty < E_{\theta_0}\log\frac{f_m(\cdot,\theta_0)}{f_m^*(\cdot,H)} \le E_{\theta_0}\log\frac{f_m(\cdot,\theta_0)}{f_m^*[\cdot,N_h(\phi)K]}.$$

It follows from the upper semi-continuity of $f(x,.)$ at ϕ that

$$f_m^*[x_1,\ldots,x_m,N_h(\phi)K] \downarrow f_m(x_1,\ldots,x_m,\phi) \text{ as } h \downarrow 0.$$

Therefore, when $h \to 0$,

$$E_{\theta_0} \log \frac{f_m(\cdot,\theta_0)}{f_m^*[\cdot,N_h(\phi)K]} \to E_{\theta_0} \log \frac{f_m(\cdot,\theta_0)}{f_m(\cdot,\phi)} > 0.$$

Hence when $h > 0$ is sufficiently small,

$$E_{\theta_0} \log\{f_m(\cdot,\theta_0)/f_m^*[\cdot,N_h(\phi)K]\} > 0,$$

and so there is an open ball $S(\phi)$ with centre ϕ such that

$$E_{\theta_0} \log \frac{f_m(\cdot,\theta_0)}{f_m^*[\cdot,S(\phi)K]} > 0.$$

$S(\phi)K$ is inferior to θ_0. Every point of K is the centre of such a ball. The set of open balls covers the compact set K, and therefore a finite subset, say $(S_1,S_2,\ldots,S_t)$ covers K. $K = \bigcup_1^t S_r K$ is inferior to θ_0.

Thus with probability one, when n is great, $f_n(x_1,\ldots,x_n,\theta)$ will attain its maximum in H at a point (or points) $\hat{\theta}_n$ in $N_{h_0}(\theta_0)$. Since h_0 can be arbitrarily small, this means that $\hat{\theta}_n \to \theta_0$ with probability one.

When condition (iv) is satisfied, the set H^c is inferior to θ_0, and so, with probability one, when n is great, the maximum in H will be a global maximum. The maximum likelihood estimator (MLE) is consistent with probability one.

□ □ □

2 Location and scale parameters. As an application of the theorem we shall consider location and scale parameters. Here $\theta = (a,c)$, $c > 0$, $\theta_0 = (a_0,c_0)$, and $f(x,\theta) = c^{-1}g[(x-a)/c]$, where g is a probability density relative to Lebesgue measure on R^1. $X_1, X_2, \ldots$ are independent random variables each with probability density $f(\cdot,\theta_0)$.

Theorem I

Let H be the compact set

$$H = \{(a,c);\ -A \le a \le A,\ c_1 \le c \le c_2\}$$

where $0 < A < \infty$, $0 < c_1 < c_2 < \infty$ *and* $\theta_0 = (a_0, c_0) \in H$. *If* g *is bounded, and upper semi-continuous, and has the property that* $K > 0$, $\lambda \geq 1$ *exist such that*

$$g(y) \leq \lambda g(x) \quad \text{if} \quad y \leq x \leq -K, \quad \text{or} \quad \text{if } K \leq x \leq y,$$

then $\hat{\theta}_n$, *the local MLE in H, for a sample of n,* $\rightarrow \theta_0$, *a.s. as* $n \rightarrow \infty$.

Proof. When $x > A + Kc_2$, $(x-a)/c \geq (x-A)/c_2 > K$,

$$f(x,\theta) = \frac{1}{c} g\left(\frac{x-a}{c}\right) \leq \frac{\lambda}{c} g\left(\frac{x-A}{c_2}\right) \leq \frac{\lambda c_2}{c_1} \frac{1}{c_2} g\left(\frac{x-A}{c_2}\right),$$

and therefore $f^*(x,H) \leq kf(x,\theta_1)$, where $k = \lambda c_2/c_1 > 1$, $\theta_1 = (A, c_2)$.

$$\int_{A+Kc_2}^{\infty} \log \frac{f(x,\theta_0)}{f^*(x,H)} f(x,\theta_0)dx \geq \int_{A+Kc_2}^{\infty} \log \frac{f(x,\theta_0)}{kf(x,\theta_1)} f(x,\theta_0)dx.$$

$$= \int_{A+Kc_2}^{\infty} \log \frac{f(x,\theta_0)}{f(x,\theta_1)} f(x,\theta_0)dx - \log k \int_{A+Kc_2}^{\infty} f(x,\theta_0)dx$$

$$> -\infty - \log k,$$

by Lemma 1 and the fact that $k > 1$.

We can show similarly that

$$\int_{-\infty}^{-A-Kc_2} \log \frac{f(x,\theta_0)}{f^*(x,H)} f(x,\theta_0)dx > -\infty.$$

Now consider

$$J = \int_{-A-Kc_2}^{A+Kc_2} \log \frac{f(x,\theta_0)}{f^*(x,H)} f(x,\theta_0)dx$$

$$= \int_{-A-Kc_2}^{A+Kc_2} \log f(x,\theta_0) f(x,\theta_0)dx - \int_{-A-Kc_2}^{A+Kc_2} \log f^*(x,H) f(x,\theta_0)dx.$$

$u \log u$ has a minimum $-e^{-1}$ at $u = e^{-1}$. Hence the first integrand $\geq -e^{-1}$. The function g is bounded, say $g(x) \leq b$.

$$f(x,\theta) = c^{-1}g[(x-a)/c] \leq b/c_1\,;\ f^*(x,H) \leq b/c_1\,.$$

Hence

$$J \geq -2(A + Kc_2)\{e^{-1} + (b/c_1)\log b/c_1\} > -\infty\,.$$

Thus

$$\int_{-\infty}^{\infty} \log\frac{f(x,\theta_0)}{f^*(x,H)} f(x,\theta_0) > -\infty\,.$$

Condition (iii) is satisfied, and the theorem is proved.

□□□

Theorem II
If the every $\alpha > 0$, $|x|^{1+\alpha}g(x) \to \infty$ as $x \to \infty$, or as $x \to -\infty$, no global MLE exists.

Proof. Suppose that, for any $\alpha > 0$,

$x^{1+\alpha}g(x) \to \infty$ as $x \to \infty$. Let $k \neq 0$ be such that $g(k) > 0$.

$$f_n(x_1,\ldots,x_n,a,c) = \frac{1}{c^n}\prod_{r=1}^{n} g\left(\frac{x_r - a}{c}\right).$$

Let x_1 be the least x_r. Put $a = x_1 - kc$.

$$f_n(x_1,\ldots,x_n,a,c) = g(k)\prod_{r=2}^{n}\frac{1}{c^{n/(n-1)}}g\left(k + \frac{x_r - x_1}{c}\right)$$

$$= g(k)\prod_{r=2}^{n}\left\{\frac{[k + (x_r - x_1)/c]^{n/(n-1)}g[k + (x_r - x_1)/c]}{(x_r - x_1 + kc)^{n/(n-1)}}\right\}$$

When $n > 1$, the factor between the braces $\to \infty$ as $c \to 0$, whether $x_r = x_1$ or not. Hence $f_n(x_1,\ldots,x_n,a,c) \to \infty$, and so no global MLE exists. □□□

Consider

$$g(x) = \frac{1}{2(1+|x|)[1+\log(1+|x|)]^2},\ -\infty < x < \infty\,.$$

For any $\alpha > 0, x^{1+\alpha}g(x) \to \infty$ when $x \to \infty$. Thus there is no global MLE for the parameters a, c. However, the conditions of Theorem II are satisfied, and the local MLE in a compact set H containing θ_0 will $\to \theta_0$ a.s. as $n \to \infty$.

Theorem III

If g is bounded and upper semi-continuous, and for some $\alpha > 0$, $|x|^{1+\alpha}g(x)$ is bounded, then with probability one, a global MLE $\hat{\theta}_n$ exists when n is great, and $\hat{\theta}_n \to \theta_0$ as $n \to \infty$.

Proof. Suppose that $g(x)$ and $|x|^{1+\alpha}g(x)$ are both $< b$. It will then be true that if $0 < m \leq 1 + \alpha$, $g(x) < b/|x|^m$.

Let x_1 be the x_r nearest to a, and let $z = \min_{r<s} |x_r - x_s|$. If $r > 1$,

$$|x_r - x_1| \leq |x_r - a| + |x_1 - a| \leq 2|x_r - a|$$

Hence

$$|x_r - a| \geq \tfrac{1}{2}z, \quad r > 1.$$

$$g\left(\frac{x_r - a}{c}\right) < \frac{bc^{n/(n-1)}}{|x_r - a|^{n/(n-1)}} \text{ if } n/(n-) < 1 + \alpha, \text{ i.e. } n > 1 + 1/\alpha,$$

$$\leq \frac{bc^{n/(n-1)}}{(\frac{1}{2}z)^{n/(n-1)}}, \quad r > 1.$$

$$f_n(x_1, \ldots, x_n, a, c) = \frac{1}{c^n} g\left(\frac{x_1 - a}{c}\right) \prod_{r=2}^{n} g\left(\frac{x_r - a}{c}\right) < \frac{2^n b^n}{z^n}.$$

Therefore

$$f_n^*(x_1, \ldots, x_n, \Theta) \leq 2^n b^n / z^n.$$

$$E_{\theta_0} \log f_n^*(X_1, \ldots, X_n, \Theta) \leq n \log(2b) - nE_{\theta_0} \log Z < \infty, \qquad [8.3]$$

because $E_{\theta_0} \log Z > -\infty$; see Lemma 4.

It is easy to show by differentiation that if $u > 0, m > 0$, then

$$\log u \geq (1 - u^{-m})/m.$$

Therefore

$$\log g(x) \geq [1 - g(x)^{-m}]/m.$$

Hence

$$\int_{-\infty}^{\infty} \log g(x) \cdot g(x) dx > -\infty \text{ if } \int_{-\infty}^{\infty} g(x)^{1-m} dx < \infty.$$

$$g(x) < \frac{b}{|x|^{1+\alpha}}, \quad g(x)^{1-m} < \frac{b^{1-m}}{|x|^{(1+\alpha)(1-m)}} \text{ if } m < 1.$$

Take $m = \frac{1}{2}\alpha/(1+\alpha)$, $(1+\alpha)(1-m) = 1+\frac{1}{2}\alpha$

$$g(x)^{1-m} < \frac{b^{1-m}}{|x|^{1+\alpha/2}}, \text{ and also } g(x)^{1-m} < b^{1-m}.$$

Hence

$$\int_{-\infty}^{\infty} g(x)^{1-m}dx < \infty, \text{ and } \int_{-\infty}^{\infty} \log g(x)\cdot g(x)dx > -\infty,$$

$$E_{\theta_0} \log f(\cdot,\theta_0) = \int_{-\infty}^{\infty} \log\left[\frac{1}{c_0}g\left(\frac{x-a_0}{c_2}\right)\right]\frac{1}{c_0}g\left(\frac{x-a_0}{c_0}\right)dx \quad [8.4]$$

$$= \int_{-\infty}^{\infty} \log[g(x)/c_0]g(x)dx > -\infty.$$

From [8.3] and [8.4] it follows that

$$E_{\theta_0} \log\frac{f_n(\cdot,\theta_0)}{f_n^*(\cdot,\Theta)} = nE_{\theta_0}\log f(\cdot,\theta_0) - E_{\theta_0}\log f_n^*(\cdot,\Theta) > -\infty.$$

Condition (iii″) is satisfied. We now need to show that (iv′) is satisfied.

$$g\left(\frac{x_r-a}{c}\right) < b\left(\frac{c}{|x_r-a|}\right)^{1+a}$$

$$\frac{1}{c}g\left(\frac{x_r-a}{c}\right) < \frac{bc^{\alpha}}{|x_r-a|^{1+\alpha}} \le \frac{bc^{\alpha}}{(\frac{1}{2}z)^{1+\alpha}} \quad \text{if } r>1.$$

$$\frac{1}{c}g\left(\frac{x_r-a}{c}\right) < \frac{b}{c},$$

$$f_n(x_1,\ldots,x_n,a,c) < \frac{b^n c^{(n-1)\alpha-1}}{(\frac{1}{2}z)^{(n-1)(1+\alpha)}}.$$

When $(n-1)\alpha > 1$, i.e. $n > 1+1/\alpha$, this will $\to 0$ uniformly with respect to a as $c\to 0$ if $z \neq 0$, i.e. except on a set of Lebesgue measure zero.

$$\frac{1}{c}g\left(\frac{x_r-a}{c}\right) < \frac{b}{c},$$

and so $\to 0$ uniformly with respect to a as $c \to \infty$.

$$\frac{1}{c} g\left(\frac{x_r - a}{c}\right) < \frac{bc}{c|x_r - a|} = \frac{b}{|x_r - a|},$$

and therefore $\to 0$ uniformly with respect to c as $|a| \to \infty$. Hence $f_n(x_1, \ldots, x_n, a, c) \to 0$ uniformly with respect to a as $c \to 0$ or ∞, and $\to 0$ uniformly with respect to c as $|a| \to \infty$. Condition (iv′) will be satisfied. We have shown above that condition (iii″) is satisfied, and so the theorem is proved.

□ □ □

Lemma 4

If $X_1, X_2, \ldots, X_n$ are i.i.d.r.v. with a bounded density function relative to Lebesgue measure on the real line, and $Z = \min_{r<s} |X_r - X_s|$, then $E \log Z > -\infty$.

Proof. If a random variable U has a bounded density h relative to Lebesgue measure on the real line,

$$\int_{-1}^{1} \log|u| h(u) du \text{ is finite.}$$

Therefore

$$E \log|U| = \int_{-\infty}^{\infty} \log|u| h(u) du > -\infty.$$

$X_r, X_s, r < s$, are independent random variables with a bounded density, and therefore $|X_r - X_s|$ has a bounded density. Hence $E \log|X_r - X_s| > -\infty$.

For a random variable W, define

$$\begin{aligned} W^- &= W, \text{ if } W < 0 \\ &= 0, \quad \text{if } W \geq 0. \end{aligned}$$

Obviously

$$E(W) > -\infty \Leftrightarrow E(W^-) > -\infty.$$

Let $W_1, \ldots, W_k$ be random variables, each with a mean value $> -\infty$.

$$\min_r W_r \geq \sum_1^k W_r^-,$$

$$E(\min_r W_r) \geq \sum_1^k E(W_r^-) > -\infty.$$

$$\log Z = \min \log|X_r - X_s|, r > s,$$

$$E \log|X_r - X_s| > -\infty, \text{ and so } E \log Z > -\infty.$$

□□□

3 Discrete probability space. Although the theorem in Section 1 of this chapter applies to all distributions, we can get a stronger and more useful result by starting afresh, and using the special properties of a discrete probability space. This investigation points to the great difference between continuous probability distributions and discrete probability distributions. From the point of view of experiment, the former are unreal. This section is based on Hannan (1960): the result obtained is a slight extension of Hannan's result.

The probability space is countable, with points $x_1, x_2, \ldots$. We take the counting measure as the dominating measure μ. Consider the family of probability measures $\{P\}$ with densities $\{p\}$, and suppose that the actual probability measure is $P_0 \in \{P\}$, with density p_0. Denote the observations by $y_1, y_2, \ldots$.

We shall consider estimating the probability measure P_0 itself rather than parameters which determine it. For a sample of n, the MLE would be the probability measure $\hat{P}_n$ which maximizes $\sum_{r=1}^{n} \log p(y_r)$. To avoid considering compactness, we shall consider a more general estimator. The probability measure P_n, with density p_n, is a practical likelihood estimator (PLE) if for some positive sequence (ε_n) such that $\varepsilon_n/n \to 0$ as $n \to \infty$,

$$\sum_{r=1}^{n} \log p_n(y_r) \geq \sup_p \left[\sum_{r=1}^{n} \log p(y_r) \right] - \varepsilon_n. \qquad [8.5]$$

In particular, ε_n might be constant. It follows from [8.5] that

$$\sum_{r=1}^{n} \log p_n(y_r) \geq \sum_{r=1}^{n} \log p_0(y_r) - \varepsilon_n,$$

$$\frac{1}{n} \sum_{r=1}^{n} \log \frac{p_0(y_r)}{p_n(y_r)} \leq \varepsilon_n/n,$$

and therefore

$$\limsup_{n\to\infty} \frac{1}{n} \sum_{r=1}^{n} \log \frac{p_0(y_r)}{p_n(y_r)} \leq 0. \qquad [8.6]$$

A sequence (P_n) of probability measures with densities (p_n) for which [8.6] is true will be called a regular likelihood estimator (RLE). Every PLE is an RLE.

The empirical probability distribution of the sample $y_1, y_2, \ldots, y_n$ will be denoted by P_n^* with density p_n^*. If h is a function on $\mathscr{X}$,

$$\sum_{r=1}^{n} h(y_r)/n = \sum_{x} p_n^*(x)h(x),$$

which we shall write as $\sum p_n^* h$. If $E_{P_0}h$ is finite, $\sum p_n^* h \to E_{P_0}h$ a.s. as $n \to \infty$, by the strong law of large numbers. Note that [8.6] may be written

$$\limsup_{n\to\infty} \sum p_n^* \log p_0/p_n \leq 0.$$

Theorem

Let P_n be an RLE. If for some value of m

$$E_{P_0} \sup_{P} \log \prod_{r=1}^{m} \frac{p(Y_r)}{p_0(Y_r)} < \infty, \qquad [8.7]$$

then, with probability one, $P_n \to P_0$, i.e. $\rho^*(P_n, P_0) \to 0$ as $n \to \infty$.

Note that the condition [8.7] is of the type (iii″); but in the case of a discrete probability space, no further condition like (iv′) is required. In fact, we ignore the topology of Θ, and consider only that of $\{P\}$.

Proof. We shall first prove the above theorem for $m = 1$. If $\gamma(x) = \sup_P p(x)$, the condition [8.7] becomes

$$E_{P_0} \log \gamma/p_0 < \infty. \qquad [8.8]$$

Note that $\gamma \leq 1$, and therefore $E_{P_0} \log \gamma/p_0 \leq -E_{P_0} \log p_0$. Hence a sufficient condition for [8.8] is $E_{P_0} \log p_0 > -\infty$.

If $k, k' > 0$,

$$\begin{aligned} k\log(k/k') &= 2k\log(\sqrt{k}/\sqrt{k'}) \geq 2k(1 - \sqrt{k'}/\sqrt{k}) \\ &= 2\sqrt{k}(\sqrt{k} - \sqrt{k'}) = (\sqrt{k} - \sqrt{k'})^2 + (k - k'). \end{aligned}$$

$$(\sqrt{k} - \sqrt{k'})^2 \leq k' - k + k\log(k/k').$$

Let J_w denote the set $(x_1, x_2, \ldots, x_w)$ of points in $\mathscr{X}$, and also its indicator function. Put $k = J_w p_n^*$, $k' = J_w p_n$. Summing over the points of $\mathscr{X}$, we have

$$\begin{aligned}\sum J_w(\sqrt{p_n^*} - \sqrt{p_n})^2 &\leq \sum J_w(p_n - p_n^*) + \sum J_w p_n^* \log (p_n^*/p_n),\\ &\leq \sum(1 - J_w)p_n^* + \sum J_w p_n^* \log p_n^*/p_0 + \sum p_n^* \log p_0/p_n\\ &\quad + \sum(1 - J_w)p_n^* \log p_n/p_0\\ &\leq \sum p_n^*(1 - J_w)[1 + \log \gamma/p_0] + \sum J_w p_n^* \log p_n^*/p_0\\ &\quad + \sum p_n^* \log p_0/p_n.\end{aligned}$$

When $n \to \infty$, the first term on the right hand side $\to E_{p_0}(1 - J_w)[1 + \log \gamma/p_0]$ with probability one, because $E_{p_0} \log \gamma/p_0$ is finite. The second term $\to 0$ almost surely, and $\limsup_{n\to\infty} \sum p_n^* \log p_0/p_n \leq 0$. Therefore

$$\limsup_{n\to\infty} \sum J_w(\sqrt{p_n^*} - \sqrt{p_n})^2 \leq E_{p_0}(1 - J_w)[1 + \log \gamma/p_0] \quad \text{a.s.}$$

Both sides of this inequality are monotone in w, and the right side $\to 0$ as $w \to \infty$, since $E_{p_0} \log \gamma/p_0$ is finite. Hence

$$\lim_{n\to\infty} \sum J_w(\sqrt{p_n^*} - \sqrt{p_n})^2 = 0 \text{ for all } w, \quad \text{a.s.}$$

and therefore

$$\lim_{n\to\infty} [p_n^*(x) - p_n(x)] = 0 \text{ for all } x, \quad \text{a.s.}$$

Also

$$p_n^*(x) \to p_0(x) \text{ for all } x, \quad \text{a.s.},$$

Therefore

$$p_n(x) \to p_0(x) \text{ for all } x, \text{ a.s.}$$

Since

$\sum p_n(x) = \sum p_0(x)$, this implies $p_n(x) \to p_0(x)$ in mean a.s., and $\sum |p_n(x) - p_0(x)| \to 0$, a.s.

Now consider the case where [8.7] is not true for $m = 1$, but is true for some $m > 1$. First suppose $n \to \infty$ by integral multiples of $m, n = vm$, and $v \to \infty$. A sample of m observations from the space $\mathscr{X}$ may be regarded as a sample of one from the product space $\mathscr{Y} = \mathscr{X}^m$, each point of which is an ordered set of m points from $\mathscr{X}$, not necessarily all different. If P is a probability measure on $\mathscr{X}$, we shall use the same symbol P for the product measure on $\mathscr{Y} = \mathscr{X}^m$. Thus if

$$z, \in \mathscr{Y}, = (y_1, y_2, \ldots, y_m), \; P(z) = \prod_1^m P(y_r).$$

$$\frac{1}{n} \log \prod_{r=1}^{n} \frac{P_0(y_r)}{P_n(y_r)} = \frac{1}{m} \times \frac{1}{v} \log \prod_{s=1}^{v} \frac{P_0(z_s)}{P_n(z_s)},$$

where

$$z_s = \{y_{(s-1)m+1}, y_{(s-1)m+2}, \ldots, y_{(s-1)m+m}\}.$$

If P_n is an RLE for the sample of $n = vm$ from $\mathscr{X}$,

$$\limsup \frac{1}{n} \log \prod_{r=1}^{n} \frac{P_0(y_r)}{P_n(y_r)} \leq 0.$$

Therefore

$$\limsup \frac{1}{v} \log \prod_{s=1}^{v} \frac{P_0(z_s)}{P_n(z_s)} \leq 0,$$

and P_n on $\mathscr{Y}$ is an RLE for the sample $(z_1, z_2, \ldots, z_v)$ from $\mathscr{Y}$.

The condition [8.7] may be written

$$E_{P_0} \sup_P \log \frac{P(Z)}{P_0(Z)} < \infty, \; Z = (Y_1, Y_2, \ldots, Y_m).$$

It follows from the part of the theorem already proved that with probability one,

$$P_n(z) \to P_0(z), \text{ all } z.$$

In particular

$$P_n(x, x, \ldots, x) \to P_0(x, x, \ldots, x), \text{ all } x,$$
$$P_n(x)^m \to P_0(x)^m.$$

Thus

$$P_n(x) \to P_0(x), \text{ for all } x, \text{ a.s.}$$

Now suppose $n = vm + u,\ 0 < u < m$.

$$\frac{1}{n}\log\prod_{r=1}^{n}\frac{P_0(y_r)}{P_n(y_r)} = \frac{1}{n}\log\prod_{r=1}^{u}\frac{P_0(y_r)}{P_n(y_r)} + \frac{v}{vm+u}\times\frac{1}{v}\log\prod_{s=1}^{v}\frac{P_0(z_s)}{P_n(z_s)}$$

$$z_s = \{y_{(s-1)v+u+1}, \ldots, y_{(s-1)v+u+m}\}.$$

For fixed u, when $n \to \infty$, the first term on the right $\to 0$, and we can show as before that if P_n is an RLE for the sample of $n = vm + u$ from $\mathscr{X}$, then P_n on $\mathscr{Z}$ is an RLE for the sample $(z_1, \ldots, z_v)$, and $P_n \to P_0$, a.s. This is true for every $u < m$, and so the theorem is proved. □□□

CHAPTER NINE

THE SAMPLE DISTRIBUTION FUNCTION

1 The distribution function F of a real random variable X is defined by $F(x) = P\{X \leq x\}$. For a sample of n values of X, the sample distribution function K_n is defined by $K_n(x) = N(x)/n$, where $N(x) =$ number of sample values $\leq x$. It is a step function. The Glivenko–Cantelli theorem (published in 1933) states that, with probability one, $\sup_x |K_n(x) - F(x)| \to 0$ as $n \to \infty$. This is the existence theorem for Statistics as a branch of Applied Mathematics.

The Glivenko–Cantelli theorem is only the beginning. It being true, we immediately want to know how likely K_n is to differ much from F, i.e. we are interested in the distribution of the random function K_n. The investigation of its probability distribution in finite samples, and of its limiting distribution, was started by Kolmogorov with a paper in the same journal and in the same year (1933). The former is a problem in combinatorial mathematics which has taken mathematicians about 40 years to solve completely.

The sort of thing we want to know about the distribution of K_n is the probability that K_n lies between two specified functions,

$$g(x) \leq K_n(x) \leq h(x), \text{ all } x.$$

The fact that K_n is a non-decreasing function makes the second inequality equivalent to

$$K_n(x) \leq h_0(x),$$

where h_0 is the greatest non-decreasing function which is $\leq h$. Hence the effective upper barrier will be non-decreasing. So we may as well take h as non-decreasing. Similarly for g.

Again, if $F(x)$ is constant over an interval, then K_n must be constant over that interval, and so the effective upper and lower barrier functions will be constant over that interval, and therefore they will be functions of F. Thus there is no loss in generality in considering only inequalities of the form

$$g[F(x)] \leq K_n(x) \leq h[F(x)], \text{ all } x, \qquad [9.1]$$

where h and h are non-decreasing, and $g(0) = 0$, $h(1) = 1$. Kolmogorov (1933) considered the probability of

$$\max[F(x) - a, 0] \leq K_n(x) \leq \min[F(x) + a, 1].$$

For any random variable X with distribution function F,

$$F(X) \leq F(x) \Leftrightarrow X \leq x \text{ or } (X > x)[F(X) = F(x)].$$

The last event is of probability 0. Hence, with probability one,

$$X \leq x \Leftrightarrow F(X) \leq F(x).$$

Therefore, if $x_1, \ldots, x_n$ is a sample of n values of X,

$$K_n(x) = (\text{number of } x_r \leq x)/n = [\text{number of } F(x_r) \leq F(x)]/n, \text{ a.s.}$$
$$= K_n^*[F(x)],$$

where K_n^* is the sample distribution function of

$$F(x_1), \ldots, F(x_n).$$

The inequality [9.1] has therefore the same probability as

$$g[F(x)] \leq K_n^*[F(x)] \leq h[F(x)], \text{ all } x. \qquad [9.2]$$

If $0 \leq k \leq 1$, and F is continuous, then for some $x_0, F(x_0) = k$. Therefore

$$P[F(x) \leq k] = P[F(x) \leq F(x_0)] = P[X \leq x_0] = F(x_0) = k.$$

Thus if X is a random variable with a continuous distribution function F, the random variable $F(X)$ has a rectangular [0, 1] distribution. The probability of [9.2], and therefore of [9.1], will be the same as the probability of

$$g(u) \leq K_n^*(u) \leq h(u), \text{ all } u, \qquad [9.3]$$

where K_n^* is the sample distribution function of a sample of n values of a rectangular [0, 1] random variable. In other words, the probability of [9.1] is the same for all continuous distributions, and is equal to the probability of

$$g(x) \leq K_n^*(x) \leq h(x), \text{ all } x, \qquad [9.4]$$

where $K_n^*(x)$ is the sample distribution function for a sample of

n values of a rectangular [0, 1] random variable. From now on we restrict attention to a random variable with a continuous distribution function.

If F is the distribution function of X, then the probability of [9.1] is the same as the probability of [9.4]. But what if the distribution function of X is not F but G? What is then the probability of [9.1]? We are interested in this problem when we use an acceptance region of the form [9.1] to test the hypothesis that the distribution function of X is F—a generalised Kolmogrov test. We want to know the power of the test at G.

Suppose that the distribution function of X is G, and that $F = f(G)$, where f is a continuous distribution function on [0, 1].

$$\begin{aligned} &P\{g[F(x)] \leq K_n(x) \leq h[F(x)]\} \\ &= P\{g(f[G(x)]) \leq K_n(x) \leq h(f[G(x)])\} \\ &= P\{g(f(x)] \leq K_n^*(x) \leq h[f(x)]\}. \end{aligned}$$

The barrier functions g and h are replaced by $g(f)$ and $h(f)$. In all cases we reduce the problem to computing the probability of [9.4] for suitable g and h. We now study the distribution of K_n, the sample distribution function of a sample of n values of a rectangular [0, 1] variable—n random points in [0, 1]. We shall call the graph of K_n the sample graph, or the sample path.

2 If $U_1 \leq U_2 \leq \ldots \leq U_n$ are the order statistics from a sample of n values of a rectangular [0, 1] variable, the inequality [9.4] is equivalent to

$$u_i \leq U_i \leq v_i, \quad 1 \leq i \leq n, \tag{9.5}$$

where $u_i = \inf\{x; h(x) \geq i/n\}$

$$v_i = \sup\{x; g(x) \leq (i-1)/n\}.$$

Note that $0 \leq u_1 \leq u_2 \leq \ldots \leq u_n \leq 1,$

$$0 \leq v_1 \leq v_2 \leq \ldots \leq v_n \leq 1.$$

Lemma

Let

$$A_1, A_2, \ldots, A_{n-1}$$

$$B_1, B_2, \ldots, B_n$$

denote events such that for any integer k, the sets

$$\{A_r, B_s; r < k, s \leq k\}, \ \{A_r, B_s; r > k, s > k\}$$

are independent, then

$$P(B_1 B_2 \dots B_n A_1 A_2 \dots A_{n-1}) = \Delta_n = \det(d_{ij}), \ 1 \leq i, \ j \leq n$$

where

$$\begin{aligned} d_{ij} &= 0 && \text{if } i > j+1, \\ &= 1 && \text{if } i = j+1, \\ &= P(B_i) && \text{if } i = j, \\ &= P(B_i B_{i+1} \dots B_j A_i^c A_{i+1}^c \dots A_{j-1}^c) && \text{if } i < j, \end{aligned}$$

and A_r^c is the complement of A_r.

Proof. Note that the conditions on the events make $B_1, B_2, \dots, B_n$ independent. The events $A_1, A_2, \dots$ are 1-dependent. Put

$$\bar{A}_r = A_1 A_2 \dots A_r, \ \bar{B}_r = B_1 B_2 \dots B_r.$$

The lemma may be proved by use of the principle of inclusion and exclusion.

$$P(\bar{B}_n A_1 A_2 \dots A_{n-1}) = P(\bar{B}_n) - \sum P(\bar{B}_n A_r^c) + \sum_{r<s} P(\bar{B}_n A_r^c A_s^c)$$

$$- \dots + (-1)^{n-1} P(\bar{B}_n A_1^c A_2^c \dots A_{n-1}^c),$$

which can be shown directly to be the expansion of Δ_n. A proof by induction is shorter, and easier to print.

Assume the lemma true for $n = \text{m}$.

$$\Delta_m = P(\bar{B}_m \bar{A}_{m-1}).$$

Consider Δ_{m+1}. The elements of its last row are all zero except

$$d_{m+1,m} = 1, \ d_{m+1,m+1} = P(B_{m+1}).$$

Therefore

$$\Delta_{m+1} = P(B_{m+1})\Delta_m - \Delta_m',$$

where Δ_m' differs from Δ_m only in having B_m in the last column of Δ_m replaced by $B_m B_{m+1} A_m^c$, which satisfies the same conditions relative to the other events appearing in Δ'_m as does B_m. Therefore

$$\Delta'_m = P(\bar{B}_m B_{m+1} A^c_m \bar{A}_{m-1}) = P(\bar{B}_{m+1}\bar{A}_{m-1}A^c_m);$$
$$P(B_{m+1})\Delta_m = P(B_{m+1})P(\bar{B}_m\bar{A}_{m-1}) = P(\bar{B}_{m+1}\bar{A}_{m-1}).$$

Thus

$$\begin{aligned}\Delta_{m+1} &= P(\bar{B}_{m+1}\bar{A}_{m-1}) - P(\bar{B}_{m+1}\bar{A}_{m-1}A^c_m)\\ &= P(\bar{B}_{m+1}\bar{A}_{m-1}A_m) = P(\bar{B}_{m+1}\bar{A}_m),\end{aligned}$$

and so the lemma is true for $n = m + 1$. It is easy to show that it is true for $n = 2$, and so it is true for all n. □□□

Corollary
Taking the case where every $P(B_i) = 1$, we obtain the following result for the 1-dependent sequence of events $A_1, A_2, \ldots$,

$$P(A_1 A_2 \ldots A_{n-1}) = \det(d_{ij}),$$

where

$$\begin{aligned} d_{ij} &= 0 && \text{if } i > j+1,\\ &= 1 && \text{if } i = j \text{ or } j+1\\ &= P(A^c_1 A^c_{i+1} \cdots A^c_{j-1}) && \text{if } i < j.\end{aligned}$$

Theorem
Let

$$0 \le u_1 \le u_2 \le \ldots \le u_n \le 1,$$
$$0 \le v_1 \le v_2 \le \ldots \le v_n \le 1,$$

be given constants such that

$$u_i < v_i,\ i = 1, 2, \ldots, n.$$

If $U_1, U_2, \ldots, U_n$ are the order statistics (in ascending order) from a sample of n independent uniform random variables with range 0 to 1,

$$P(u_i \le U_i \le v_i,\ 1 \le i \le n) = n!\det[(v_i - u_j)_+^{j-i+1}/(j-i+1)!],$$

where $(x)_+ = \max(x, 0)$, and it is understood that determinant elements for which $i > j+1$ are all zero.

Proof. Let $Y_1, Y_2, \ldots, Y_n$ be independent random variables, each with a uniform distribution from 0 to 1. The required probability is equal to

$$n!P(u_i \leq Y_i \leq v_i,\ 1 \leq i \leq n;\ Y_1 \leq Y_2 \leq \ldots \leq Y_n). \quad [9.6]$$

Denote by B_i the event $u_i \leq Y_i \leq v_i$. Denote by A_i the event $Y_i \leq Y_{i+1}$, and by A_i^c the complement of A_i, that is, the event $Y_i > Y_{i+1}$. The events A_i, B_i satisfy the conditions of the lemma. Hence

$$\begin{aligned} P(u_i \leq Y_i \leq v_i,\ 1 \leq i \leq n;\ Y_1 \leq Y_2 \leq \ldots \leq Y_n) & \quad [9.7] \\ = P(B_1 B_2 \ldots B_n A_1 A_2 \ldots A_{n-1}) & \\ = \det(d_{ij}). & \end{aligned}$$

$$d_{ii} = P(B_i) = v_i - u_i.$$

If $i < j$,

$$d_{ij} = P(B_i B_{i+1} \ldots B_j A_i^c A_{i+1}^c \ldots A_{j-1}^c).$$

The event $B_i B_{i+1} \ldots B_j A_i^c A_{i+1}^c \ldots A_{j-1}^c$ is

$$u_r \leq Y_r \leq v_r,\ i \leq r \leq j$$

$$Y_i > Y_{i+1} > \ldots > Y_j.$$

This is equivalent to

$$v_i \geq Y_i > Y_{i+1} > \ldots > Y_j \geq u_j,$$

the probability of which is $(v_i - u_j)_+^{j-i+1}/(j-i+1)!$. The theorem then follows from [9.6] and [9.7]. □□□

This result is due to Steck (1971). It is an explicit solution of the problem for finite n; but it is not of great practical value except when n is small. The expansion of the determinant has 2^{n-1} non-vanishing terms. Simpler and more useful expressions for crossing probabilities can be obtained when the barriers are straight lines.

3 Probability of crossing a straight line through the origin. Let OE be the line $x = ky, 0 < k < 1$, and let p be the probability that the sample path crosses OE.

$$1 - p = P\{U_r > kr/n;\ 1 \leq r \leq n\}.$$

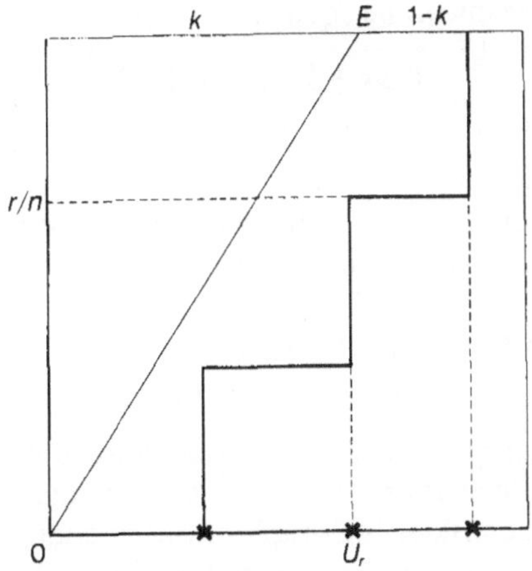

Figure 1

Put $\lambda = k/n$.

$$1 - p = n! \int_{n\lambda}^{1} du_n \ldots \int_{2\lambda}^{u_3} du_2 \int_{\lambda}^{u_2} du_1 .$$

Consider

$$Q_r = \int_{r\lambda}^{u_{r+1}} du_r \ldots \int_{2\lambda}^{u_3} du_2 \int_{\lambda}^{u_2} du_1 .$$

This will be a polynomial in u_{r+1} which vanishes when $u_{r+1} = r\lambda$

$$Q_1 = u_2 - \lambda, \; Q_2 = u_3^2/2 - \lambda u_3$$

and it is easy to show by induction that

$$Q_r = \frac{u_{r+1}^r}{r!} - \lambda \frac{u_{r+1}^{r-1}}{(r-1)!}$$

$$1 - p = n! Q_n \text{ with } u_{n+1} = 1,$$
$$= 1 - n\lambda = 1 - k.$$

Thus $p = k$. This is a very simple and elegant result. It is remarkable that the probability of crossing is independent of the sample number n. The following proof, suggested by Dr J.W. Pitman shows why this is so.

Extend the domain of definition of K_n, the sample distribution function, by means of the equation

$$K_n(x+1) = K_n(x) + 1.$$

For any $x_0 \in [0, 1]$, consider the cyclical transformation of the interval $[0, 1]$, $x \to x^*$, $x^* = x - x_0$ if $x \geq x_0$, $x^* = 1 + x - x_0$ if $x < x_0$. The sample $x_1, \ldots, x_n$ transforms to the sample $x_1^*, \ldots, x_n^*$, and the new sample distribution function K_n^* will be given by

$$\begin{aligned} K_n^*(x) &= K_n(x) - K_n(x_0), \text{ If } x \geq x_0, \\ &= K_n(x+1) - K_n(x_0), \text{ if } x < x_0. \end{aligned}$$

The set of samples which can be transformed into one another by cyclic transformations will be called a configuration. Given a configuration, consider the graph of the extended sample distribution function of a particular member sample. We may look on it as an indefinitely extended flight of steps with horizontal treads and vertical risers. Imagine that there are rays of light shining down parallel to the line $x = ky$. Each riser will cast a shadow

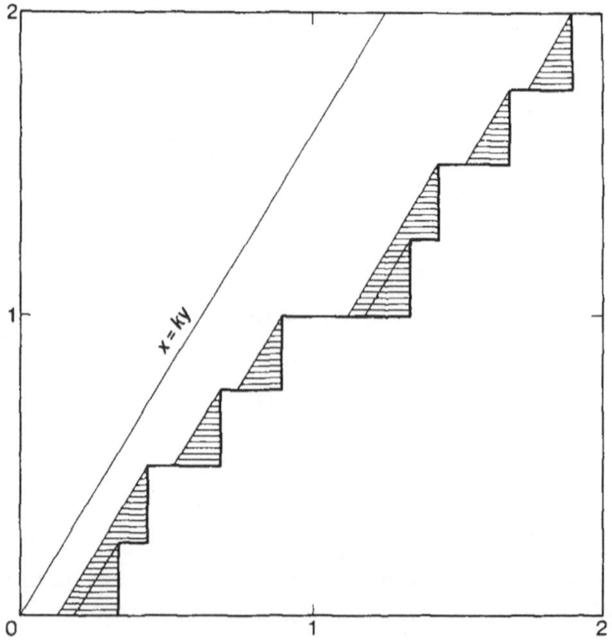

Figure 2

on one or more of the treads below, and the total width of the shadow will be k times the height of the riser. The total width of the shadows on the treads from $x=0$ to $x=1$ is therefore $kK_n(1)=k$.

The conditional distribution of the members, given the configuration, may be specified by saying that the starting point x_0 (the point that goes to 0 in the transformation) has a uniform distribution over (0, 1). Hence the probability that it is in shadow is k. The sample path of a member of the configuration crosses the line $x = ky$ if and only if its starting point is in shadow, and so the conditional probability of crossing, given the configuration, is k. This is true for all configurations, and therefore the probability of crossing is k. □□□

It should be noted that the proof has not assumed that the risers are all equal in height, and, of course, we may consider a basic interval of any length. The most general form of the result may be stated as follows:

Theorem

Let $h_1, \ldots, h_n$ be positive constants with sum h, and let $X_1, \ldots, X_n$ be independent rectangular $[0, a]$ variables. Define the sample function K_n for $0 \le x \le a$ by

$$K_n(x) = \sum_{X_r \le x} h_r.$$

Call the graph of K_n the sample path. The probability that the sample path cuts the straight line joining the origin to the point (k, a), $0 < k < a$ is k/a. □□□

4 Probability of crossing a straight line not through the origin. Let A be the point $(0, a)$, and A' the point $(1, 1 + a')$. AA' is the line $y = a + (1 + a' - a)x$. The probability that the sample path crosses AA' is $\sum_1^n t_r$, where t_r is the probability that the sample path crosses AA' *last* at $y = r/n$. Q is the point $(p_r, r/n)$.

$$r/n = a + (1 + a' - a)p_r.$$

$$p_r = \frac{r/n - a}{1 + a' - a}, \quad 1 - p_r = \frac{1 + a' - r/n}{1 + a' - a},$$

$$c = BO' = a'/(1 + a' - a).$$

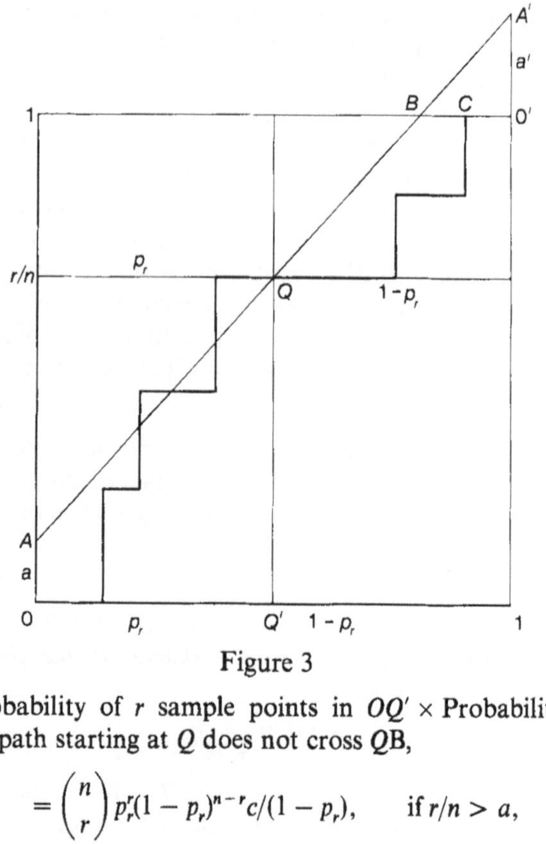

Figure 3

t_r = Probability of r sample points in OQ' × Probability that sample path starting at Q does not cross QB,

$$= \binom{n}{r} p_r^r (1-p_r)^{n-r} c/(1-p_r), \qquad \text{if } r/n > a,$$

$$= 0 \qquad \text{if } r/n \leq a,$$

The required probability of crossing is

$$P(a, a') = \frac{a'}{(1+a'-a)^n} \sum_{r>na} \binom{n}{r} (r/n - a)^r (1 + a' - r/n)^{n-r-1}.$$

Abel's formula [A.4] in Section 5 of the Appendix is

$$(z+u)^n = \sum_{r=0}^{n} \binom{n}{r} u(z-n+r)^r (u+n-r)^{n-r-1}.$$

Putting $z = n(1-a')$, $u = na'$, we obtain

$$1 = \frac{a'}{(1+a'-a)^n} \sum_{r=0}^{n} \binom{n}{r} (r/n - a)^r (1 + a' - r/n)^{n-r-1}.$$

Thus

$$P(a,a') = 1 - \frac{a'}{(1+a'-a)^n} \sum_{r<na} \binom{n}{r} (r/n - a)^r (1 + a' - r/n)^{n-r-1},$$

t_r being 0 if $r = na$.

By considering the transformation of the unit square,

$$(x, y) \to (1 - x, 1 - y),$$

which rotates Fig. 3 through 180°, we can see that

$$P(a, a') = P(-a', -a),$$

the probability of crossing the straight line joining $(0, -a')$ and $(1, 1-a)$.

Putting $a = \alpha n^{-1/2}$, $a' = \beta n^{-1/2}$, we obtain

$$P(\alpha n^{-1/2}, \beta n^{-1/2}) = \sum_{r > \alpha n^{1/2}} t_r,$$

where

$$t_r = \frac{\beta n^{-1/2}}{[1 + (\beta - \alpha) n^{-1/2}]^n} \binom{n}{r} (r/n - \alpha n^{-1/2})^r \times (1 + \beta n^{-1/2} - r/n)^{n-r-1}$$

$$= \frac{n! r^r (n-r)^{n-r-1}}{n^{n-1/2} r! (n-r)!} \times \frac{\beta (1 - \alpha n^{1/2}/r)^r (1 + \beta n^{1/2}/(n-r))^{n-r-1}}{[1 + (\beta - \alpha) n^{-1/2}]^n}$$

$$= U_r \times V_r.$$

By the use of Stirling's Theorem, we obtain

$$\log U_r = \log \frac{n}{\sqrt{[2\pi r (n-r)^3]}} + \frac{\theta_1}{12n} - \frac{\theta_2}{12r} - \frac{\theta_3}{12(n-r)},$$

for $0 < r < n$, where $0 < \theta_1, \theta_2, \theta_3 < 1$.

Therefore

$$U_r \sim \frac{n}{\sqrt{[2\pi r (n-r)^3]}}, \quad r, n - r \to \infty;$$

and

$$U_r < \frac{Cn}{\sqrt{[2\pi r (n-r)^3]}}, \qquad 0 < r < n$$

where C is a constant.

$$\log V_r = \log\beta - \log[1+\beta n^{1/2}/(n-r)]$$
$$+r\left[-\frac{\alpha n^{1/2}}{r}-\frac{\alpha^2 n}{2r^2}-\frac{\alpha^3 n^{3/2}}{3r^3}-\ldots\right]$$
$$+(n-r)\left[\frac{\beta n^{1/2}}{n-r}-\frac{\beta^2 n}{2(n-r)^2}+\frac{\beta^3 n^{3/2}}{3(n-r)^3}-\ldots\right]$$
$$-n\left[\frac{\beta-\alpha}{n^{1/2}}-\frac{(\beta-\alpha)^2}{2n}+\frac{(\beta-\alpha)^3}{3n^{3/2}}-\ldots\right]$$
$$=\log\beta-\tfrac{1}{2}\left[\frac{n\alpha^2}{r}+\frac{n\beta^2}{n-r}-(\beta-\alpha)^2\right]+\eta,$$
$$=\log\beta-\tfrac{1}{2}\frac{[(n-r)\alpha+r\beta]^2}{r(n-r)}+\eta.$$

where

$$|\eta|<\frac{\alpha^3 n^{3/2}}{r^2}+\frac{\beta^3 n^{3/2}}{(n-r)^2}+\frac{|\beta-\alpha|^3}{n^{1/2}}+\frac{\beta n^{1/2}}{n-r}$$

when $r, n-r$ great.

$$|\eta|<Kn^{-1/6} \text{ if } n^{5/6}<r<n-n^{5/6}.$$

Hence if $n^{5/6}<r<n-n^{5/6}$,

$$V_r \sim \beta\exp\left\{\frac{-\frac{1}{2}[(n-r)\alpha+r\beta]^2}{r(n-r)}\right\},\quad n\to\infty,$$

uniformly with respect to r, and

$$t_r \sim \frac{n\beta}{\sqrt{[2\pi r(n-r)^3]}}\exp\left\{\frac{-\frac{1}{2}[(n-r)\alpha+r\beta]^2}{r(n-r)}\right\}$$

uniformly with respect to r. Thus if $n^{-1/6}<r/n<1-n^{-1/6}$,

$$t_r \sim \frac{\beta}{n\sqrt{[2\pi(r/n)(1-r/n)^3]}}\exp\left\{\frac{-\frac{1}{2}[(1-r/n)\alpha+(r/n)\beta]^2}{(r/n)(1-r/n)}\right\}.$$

uniformly with respect to r. Therefore

$$\sum_{n^{5/6}}^{n-n^{5/6}} t_r \to \int_{n^{-1/6}}^{1-n^{-1/6}} \frac{\beta}{\sqrt{(2\pi)}}\frac{\exp\left\{\frac{-\frac{1}{2}[\alpha(1-x)+\beta x]^2}{x(1-x)}\right\}}{\sqrt{[x(1-x)}}dx$$

$$\to \int_0^1 \frac{\beta}{\sqrt{(2\pi)}} \frac{\exp\left\{\dfrac{-\frac{1}{2}[\alpha(1-x)+\beta x]^2}{x(1-x)}\right\}}{\sqrt{[x(1-x)^3]}} dx \qquad [9.8]$$

$$= e^{-2\alpha\beta}, \qquad [9.9]$$

see below. Using

$$\log(1+x) \geq x - x^2, \ |x| \text{ sufficiently small}$$

$$\log(1+x) \leq x - \log(1+x^2/6), \ x > -1$$

we obtain

$$\begin{aligned} \log V_r &< \log\beta + r\left(-\frac{\alpha n^{1/2}}{r}\right) \\ &\quad + (n-r)\left\{\frac{\beta n^{1/2}}{n-r} - \log\left[1+\frac{\beta^2 n}{6(n-r)^2}\right]\right\} \\ &\quad - n\left[\frac{\beta - a}{n^{1/2}} - \frac{(\beta-\alpha)^2}{n}\right] \\ &= \log\beta + (\beta-\alpha)^2 - (n-r)\log\left(1+\frac{\beta^2 n}{6(n-r)^2}\right) \end{aligned}$$

when n is great. Therefore

$$\begin{aligned} t_r = U_r V_r &< \frac{Kn}{\sqrt{[2\pi r(n-r)^3]}} \frac{1}{[1+\beta^2 n/6(n-r)^2]^{n-r}} \\ &< \frac{Kn}{\sqrt{[2\pi r(n-r)^3]}} \frac{1}{\beta^2 n/6(n-r)} = \frac{K'}{\sqrt{[r(n-r)]}}, \end{aligned}$$

where K, K' are constants.
Hence

$$\begin{aligned} \sum_{\alpha n^{1/2}}^{n^{5/6}} t_r &< K' \sum_{r=1}^{n^{5/6}} \frac{1}{n\sqrt{[(r/n)(1-r/n)]}} \\ &\to \int_0^{n^{-1/6}} \frac{dx}{\sqrt{[x(1-x)]}} \to 0 \end{aligned}$$

Similarly

$$\sum_{n-n^{5/6}}^{n} t_r \to 0,$$

and so

$$P(\alpha n^{-1/2}, \beta n^{-1/2}) = \sum_{r>\alpha n^{1/2}} t_r \to e^{-2\alpha\beta}.$$

Proof of [9.9]. By the substitution

$$u = x^{1/2}(1-x)^{-1/2}, \; du = \tfrac{1}{2}x^{-1/2}(1-x)^{-3/2}dx,$$

the integral [9.8] becomes

$$\frac{2\beta}{\sqrt{(2\pi)}} \int_0^\infty \exp[-\tfrac{1}{2}(\alpha/u + \beta u)^2]du. \qquad [9.10]$$

By the substitution

$$y = \beta u - \alpha/u, \; u = \frac{y + \sqrt{(y^2 + 4\alpha\beta)}}{2\beta}, \; du = \frac{dy}{2\beta}\left[1 + \frac{y}{\sqrt{(y^2 + 4\alpha\beta)}}\right],$$

the integral [9.10] becomes

$$\frac{1}{\sqrt{(2\pi)}} \int_{-\infty}^\infty \exp[-\tfrac{1}{2}(y^2 + 4\alpha\beta)]\left[1 + \frac{y}{\sqrt{(y^2 + 4\alpha\beta)}}\right]dy$$

$$= \frac{1}{\sqrt{(2\pi)}} \int_{-\infty}^\infty \exp[-\tfrac{1}{2}(y^2 + 4\alpha\beta)]dy = e^{-2\alpha\beta}.$$

This completes the proof that

$$\lim_{n\to\infty} P(\alpha n^{-1/2}, \beta n^{-1/2}) = e^{-2\alpha\beta}, \qquad \alpha, \beta > 0.$$

Since

$$P(-\alpha n^{-1/2}, -\beta n^{-1/2}) = P(\beta n^{-1/2}, \alpha n^{-1/2}),$$

we have also

$$\lim_{n\to\infty} P(-\alpha n^{-1/2}, -\beta n^{-1/2}) = e^{-2\alpha\beta}. \qquad \square\square\square$$

The proof for this elegant result is lengthy; but it seems the best that can be done by elementary methods. A much shorter, but more sophisticated proof depends upon the properties of the Brownian bridge. See Billingsley (1968).

When the line AA' is parallel to OO', $\beta = \alpha$, and

$$P(\alpha n^{-1/2}, \alpha n^{-1/2}) \to e^{-2\alpha^2}.$$

AA' is then the line $y = \alpha n^{-1/2} + x$. The probability of not crossing for $\alpha > 0$, is $P[K_n(x) \leq \alpha n^{-1/2} + x]$, which $\rightarrow 1 - e^{-2\alpha^2}$. For the rectangular $[0, 1]$ distribution $F(x) = x$, $0 \leq x \leq 1$. Hence, from the theory in Section 1 above, if K_n is now the sample distribution function of a sample of n values of a random variable with a continuous distribution function F,

$$P[K_n(x) - F(x) \leq \alpha n^{-1/2}] = 1 - P(\alpha n^{-1/2}, \alpha n^{-1/2}) \rightarrow 1 - e^{-2\alpha^2}.$$

5 Boundary consisting of two parallel lines.
We now consider a pair of parallel boundary lines

$$y = a + (1 + c)x, \qquad y = -b + (1 + c)x,$$

where $a, b, a + c, b - c > 0$, $a, b - c < 1$.
Let $h(\gamma)$ = probability of crossing the line $y = \gamma + (1 + c)x$, then

$h(a)$ = probability of crossing $y = a + (1 + c)x$, $= P(a, a + c)$,

$h(-b)$ = probability of crossing $y = -b + (1 + c)x$,

$= P(-b, c - b)$

$= P(b - c, b) = h(b - c)$.

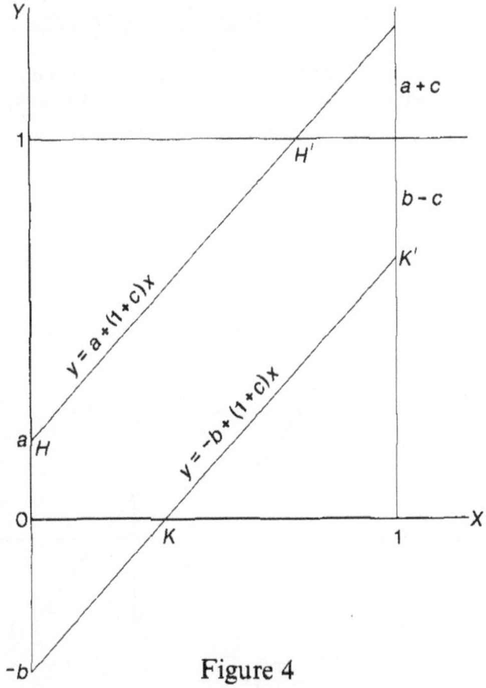

Figure 4

While various values of γ are considered, c remains fixed throughout discussion. Put

$$G(x) = N(x)/n - (1 + c)x,$$

then

$$G(0) = 0, \ G(1) = - c,$$

and

$$h(\gamma) = P[G(x) = \gamma \text{ for some } x].$$

We require the probability that the sample path crosses the boundary consisting of the two lines, HH' and KK', that is the probability that $G(x)$ takes at least one of the values $a, -b$. Denote the event, $G(x)$ takes the value a at some point in $[0, 1]$, by $\mathscr{A}$, and the event, $G(x) = -b$ at some point in $[0, 1]$, by $\mathscr{B}$. The required probability is

$$\begin{aligned} P(\mathscr{A} + \mathscr{B}) &= P(\mathscr{A}) + P(\mathscr{B}) - P(\mathscr{A}\mathscr{B}) \\ &= P(\mathscr{A}) + P(\mathscr{B}) - P(\mathscr{A} \text{ followed by } \mathscr{B}) \\ &\quad - P(\mathscr{B} \text{ followed by } \mathscr{A}). \end{aligned}$$

As x increases from 0 to 1, $G(x)$ steadily decreases except at sample points, at each of which it increases by a jump of $1/n$. The replacement of the sample points in $(t, 1)$, $0 < t < 1$, by their

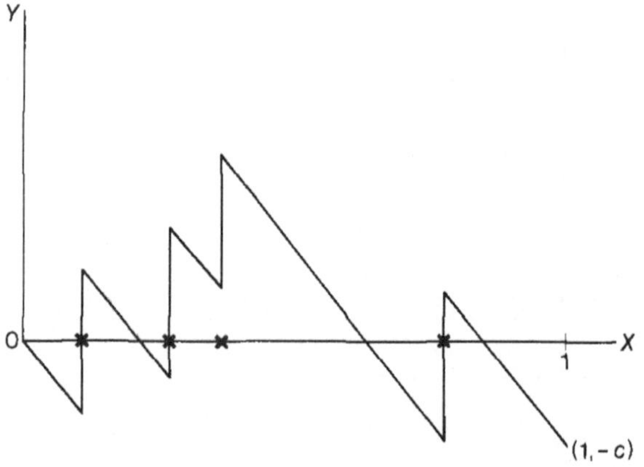

Figure 5

transforms under the transformation $x \to t+1-x$, will be called reversal after t. Suppose $G(t_1)=k_1, G(t_2)=k_2$, where $0<t_1<t_2<1$. As x increases from t_2 to 1, $G(x)$ changes from k_2 to $-c$, a *decrease* (positive or negative) of k_2+c. If we reverse the sample points after $t_1, G(x)$ will decrease by k_2+c as x increases from t_1 to t_1+1-t_2, and so will take the value k_1-k_2-c at t_1+1-t_2.

A sample in which $G(x)$ takes the value γ at some x, will be called a γ sample. When $\gamma \geq a$, such a sample will be called a weak γ sample, $W(\gamma)$, if $G(x)$ takes the value $-b$ before the value a; otherwise the sample is strong, $S(\gamma)$. If $\gamma \leq -b$, a γ sample is weak if $G(x)$ takes the value a before the value $-b$: otherwise it is strong. The probability of crossing the boundary is

$$h(a)+h(b-c)-P[W(a)]-P[W(-b)].$$

Consider a sample which is $W(A), A \geq a$, and let t be the point at which $G(x)$ first takes the value $-b$. At some point in $(t, 1)$, $G(x)$ takes the value A. Hence by reversal after t, the sample will become a $-b-A-c$ sample. Moreover, it will become a strong $-b-A-c$ sample. By this process, every $W(A)$ will become an $S(-b-A-c)$. By the same process, of reversal after the point at which $G(x)$ first takes the value $-b$, every $S(-b-A-c)$ will become $W(A)$, and the mapping is one-to-one. Hence

$$\begin{aligned} P[W(A)] &= P[S(-A-b-c)] \\ &= h(-A-b-c)-P[W(-A-b-c)] \\ &= h(A+b)-P[W(-A-b-c)]. \qquad [9.11] \end{aligned}$$

Now consider a $W(-B), B \geq b$. Let t be the point where $G(x)$ is first $\geq a$, then $G(t)=a+d$, where $0 \leq d < 1/n$, ignoring throughout the whole discussion the possibility, of zero probability, that any two sample points coincide. By reversal after t, the sample will become $S(a+d+B-c)$, which is included in the set of $S(B+a-c)$. Therefore

$$\begin{aligned} P[W(-B) \leq P[S(B+a-c)] &= h(B+a-c) \\ &\quad -P[W(B+a-c)]. \qquad [9.12] \end{aligned}$$

By the same process, an $S(B+a-c)$ will become $W(-B+d)$. Therefore an $S(B+a-c+1/n)$ will become $W(-B+d-1/n)$, which is also $W(-B)$, since $d - 1/n \leq 0$. Thus

$$P[W(-B)] \geq P[S(B + a - c + 1/n)]$$
$$= h(B + a - c + 1/n) - P[W(B + a - c + 1/n)]. \qquad [9.13]$$

Combining inequality [9.11] with either [9.12] or [9.13], we obtain

$$P[W(A)] \geq h(A + b) - h(A + b + a) + P[W(A + b + a)], \quad A \geq a,$$
$$\leq h(A + b) - h(A + b + a + 1/n)$$
$$+ P[W(A + b + a + 1/n)].$$
$$P[W(-B) \geq h(B + a - c + 1/n) - h(B + a - c + 1/n)$$
$$+ P[W(-B - a - b - 1/n)], \quad B \geq b,$$
$$\leq h(B + a - c) - h(B + a + b - c)$$
$$+ P[W(-B - a - b)].$$

Each of these can be extended indefinitely by repeatedly using the same inequality on the last term. From these we deduce

$$P[W(a)] \geq h(a + b) - h(2a + b) + h(2a + 2b) - \ldots$$
$$\leq h(a + b) - h(2a + b + 1/n)$$
$$+ h(2a + 2b + 1/n) - h(3a + 2b + 2/n) + \ldots$$
$$P[W(-b)] \geq h(a + b - c + 1/n) - h(a + 2b - c + 1/n)$$
$$+ h(2a + 2b - c + 2/n) - \ldots$$
$$\leq h(a + b - c) - h(a + 2b - c)$$
$$+ h(2a + 2b - c) - \ldots$$

The terms on the right hand sides of the inequalities decrease in magnitude rapidly. Note that

$$P[W(a)] + P[W(-b)] \leq h(a + b) + h(a + b - c)$$
$$\geq h(a + b) + h(a + b - c + 1/n)$$
$$- h(2a + b) - h(a + 2b - c),$$

which shows that $h(a + b) + h(a + b - c)$ is a very good approximation to $P[W(a)] + P[W(-b)]$. Thus the probability that the sample path crosses the boundary consisting of the two parallel lines $y = a + (1 + c)x$, $y = -b + (1 + c)x$ is approximately

$$h(a) + h(-b) - h(a + b) - h(a + b - c),$$

$$= P(a, a + c) + P(b - c, b) - P(a + b, a + b + c)$$

$$- P(a + b - c, a + b).$$

The most important case is $c = 0$, the boundary lines parallel to the line $y = x$. Then $h(\gamma) = P(\gamma, \gamma)$, and when $n \to \infty, h(\gamma n^{-1/2})$ and $h(\gamma n^{-1/2} + n^{-1})$ both $\to e^{-2\gamma^2}$. Therefore the probability of crossing the boundary consisting of the lines $y = \pm \gamma n^{-1/2} + x$

$$= 2h(\gamma n^{-1/2}) - 2P[W(\gamma n^{-1/2})]$$

$$\to 2 \sum_{r=1}^{\infty} (-1)^{r-1} e^{-2r^2\gamma^2}.$$

Now let K_n be the sample distribution function of a sample of n values of a random variable with a continuous distribution function F, and let $D_n = \sup_x |K_n(x) - F(x)|$, then

$$P\{D_n \geq \gamma n^{-1/2}\} = 2h(\gamma n^{-1/2}) - 2P[W(\gamma n^{-1/2})]$$

$$\to 2[e^{-2\gamma^2} - e^{-8\gamma^2} + e^{-18\gamma^2} - \ldots], \text{ as } n \to \infty.$$

The second term in the series between the braces is the fourth power of the first term. It, and all subsequent terms, can usually be neglected in practical applications. Inequalities similar to those in this section are discussed in Durbin (1968).

APPENDIX MATHEMATICAL PRELIMINARIES

1 Convergence in mean and convergence in measure. A sequence (f_n) of real valued integrable functions on a space $\mathscr{X}$ with a measure μ is said to converge in mean to f if $\lim \int |f_n - f| = 0$. It then follows that $\int f_n \to \int f$, and that $|f_n|$ converges to $|f|$ in mean. The sequence (f_n) converges in measure to f if for every $\varepsilon > 0$,

$$\mu\{x; x \in \mathscr{X}, |f_n(x) - f(x)| > \varepsilon\} \to 0 \text{ as } n \to \infty.$$

The sequence (f_n) may converge to f in mean, or in measure, without converging almost everywhere to f; but it is well known that in both cases every subsequence of (f_n) contains a subsequence which converges to f almost everywhere. It is convenient to have a name for this type of point convergence: we shall call it *loose convergence.* We shall say that g_n *converges loosely* to g, and write $g_n \xrightarrow{l} g$, if every subsequence of (g_n) contains a subsequence which converges almost everywhere to g. It is easy to show that if $\mu(\mathscr{X}) < \infty$, loose convergence implies convergence in measure.

Loose convergence obeys the usual manipulative rules of point convergence, such as

$$f_n \xrightarrow{l} f, \; g_n \xrightarrow{l} g, \; \Rightarrow f_n + g_n \xrightarrow{l} f + g, \; f_n g_n \xrightarrow{l} fg.$$

We need a modification of Fatou's lemma.

Lemma *If* $g_n \geq 0$, *and* $\xrightarrow{l} g$, *then* $\liminf_{n \to \infty} \int g_n \geq \int g$.

Proof. Put $L = \liminf_{n \to \infty} \int g_n$. There is a sequence (n') of positive integers such that $\int g_{n'} \to L$. This sequence contains a subsequence (n'') such that $g_{n''} \to g$ almost everywhere and $\int g_{n''} \to L$. Hence, by Fatou's lemma,

$$L = \lim \int g_{n''} \geq \int g. \qquad \square\square\square$$

We shall make frequent use of the following extension of the dominated convergence theorem, which is not so widely known as it ought to be. It is essentially given in Pratt (1960), though not quite in the form given here. It does not appear in most textbooks on measure and integration. An exception is Royden (1968), but the full implications are not set out there.

Theorem

(i) $g_n \xrightarrow{l} g$, $|g_n| \leq |G_n|$ *a.e.*, G_n *integrable and* $\to G$ *in mean* $\Rightarrow g_n \to g$ *in mean.*

(ii) $H_n \geq 0$ *and integrable,* $H_n \xrightarrow{l} H$ *integrable,* $\int H_n \to \int H$ $\Rightarrow H_n \to H$ *in mean.*

Proof. We first prove $g_n \xrightarrow{l} g$, $|g_n| \leq H_n$ a.e., H_n integrable and $\xrightarrow{l} H$ integrable, $\int H_n \to \int H \Rightarrow g_n \to g$ in mean.

$$H_n + H - |g_n - g| \geq H_n + H - |g_n| - |g| \geq 0, \text{ a.e.}$$

$$H_n + H - |g_n - g| \xrightarrow{l} 2H.$$

Therefore by the Lemma,

$$\begin{aligned}\int 2H &\leq \liminf \int (H_n + H - |g_n - g|)\\ &\leq \limsup \int (H_n + H - |g_n - g|)\\ &\leq \limsup \int (H_n + H) = \int 2H.\end{aligned}$$

Hence

$$\lim \int (H_n + H - |g_n - g|) = \int 2H = \lim \int (H_n + H).$$

Thus

$$\lim \int |g_n - g| = 0, \; g_n \to g \text{ in mean.}$$

Since $G_n \to G$ in mean $\Rightarrow |G_n| \to |G|$ in mean, putting $H_n = |G_n|$, we obtain (i). Putting $g_n = H_n$, we obtain (ii).

□□□

Corollary

$$\int g_n^2 \to \int g_0^2, \; g_n \xrightarrow{l} g_0 \Leftrightarrow \int (g_n - g_0)^2 \to 0.$$

Proof. Suppose the left-hand statement true. $g_n^2 \xrightarrow{l} g_0^2$ and so by (ii) $g_n^2 \to g_0^2$ in mean. $(g_n - g_0)^2 \xrightarrow{l} 0$, and $(g_n - g_0)^2 \leq 2g_n^2 + 2g_0^2$, which $\to 4g_0^2$ in mean. Therefore $(g_n - g_0)^2 \to 0$ in mean, $\int(g_n - g_0)^2 \to 0$, $g_n \to g_0$ in quadratic mean.

If the right-hand statement is true, i.e. $(g_n - g_0)^2 \to 0$ in mean, then $g_n \overset{l}{\to} g_0$, $g_n^2 \overset{l}{\to} g_0^2$.

$$g_n^2 = (g_n - g_0 + g_0)^2 \leq 2(g_n - g_0)^2 + 2g_0^2,$$

which is convergent in mean. Therefore $g_n^2 \to g_0^2$ in mean. □□□

If f_n and f_0 are probability density functions, $\int f_n = \int f_0 = 1$, and so if $f_n \overset{l}{\to} f_0$, then by (ii) $f_n \to f_0$ in mean, and $\int |f_n - f_0| \to 0$. This is Scheffé's theorem.

A simple example of the application of this theorem (in many cases all that is required) is the following.

$F_n \geq 0$, integrable, and $\to F$ in mean; $G_n \geq 0$, integrable, and $\to G$ in mean $\Rightarrow \sqrt{(F_n G_n)} \to \sqrt{(FG)}$ in mean.

Proof. $F_n \overset{l}{\to} F$, $G_n \overset{l}{\to} G$; therefore $\sqrt{(F_n G_n)} \overset{l}{\to} \sqrt{(FG)}$. Also $\sqrt{(F_n G_n)} \leq F_n + G_n$, which $\to F + G$ in mean.

2 Mappings of measure spaces. Let μ be a σ-finite measure on a σ-algebra $\mathscr{F}$ of sets in a space $\mathscr{X}$. T is a mapping from $\mathscr{X}$ into a space $\mathscr{T}$. ν_0 is the measure induced in $\mathscr{T}$ on the σ-algebra $\mathscr{A}$; i.e. $\mathscr{A}$ is the σ-algebra of sets A in $\mathscr{T}$ such that $T^{-1}A \in \mathscr{F}$, and $\nu_0(A) = \mu(T^{-1}A)$. We shall assume that the single point sets of $\mathscr{T}$ are $\mathscr{A}$ measurable, so that the mapping partitions $\mathscr{X}$ into $\mathscr{F}$ measurable sets $T^{-1}\{t\}$, each of which is the inverse image of a single point t in $\mathscr{T}$.

Let ν be a σ-finite measure on $\mathscr{A}$ which dominates ν_0. There always is such a measure ν, because μ, being σ-finite, is dominated by some finite measure μ_1, say, and the measure induced in $\mathscr{T}$ from μ_1 is finite and dominates ν_0. If ν_0 is σ-finite, we may take $\nu = \nu_0$; but this is sometimes not so. For example, if μ is Legesgue measure in R^2 and T is the mapping into R^1 defined by $(x, y) \to x$, the induced measure ν_0 in R^1 takes only the values 0 and ∞, and so is not σ-finite. However, ν_0 is dominated by the Lebesgue measure on R^1, and we take this for ν.

Let f be a real-valued measurable function on $\mathscr{X}$ which is integrable. Put

$$Q(A) = \int_{T^{-1}A} f d\mu.$$

$$\nu(A) = 0 \Rightarrow \mu(T^{-1}A) = 0 \Rightarrow Q(A) = 0.$$

Hence $\nu \gg Q$, and so by the Radon–Nikodym theorem there exists a function g on $\mathscr{T}$, determined up to ν equivalence, such that

$$\int_{T^{-1}A} f d\mu = Q(A) = \int_A g d\nu,$$

for every $A \in \mathscr{A}$.

We shall write $g = T^*f$. There is evidently some connection with conditional expectation, and, in fact, if μ is a probability measure, and ν the induced probability measure in $\mathscr{T}$,

$$E\{f \mid T\} = g(T).$$

The mapping T^* of integrable functions on $\mathscr{X}$ into integrable functions on $\mathscr{T}$ is linear.

(i) $T^*(c_1 f_1 + c_2 f_2) = c_1 T^* f_1 + c_2 T^* f_2$, a.e.$\nu$,
for constants c_1, c_2. It is also sign-preserving.

(ii) $f \geq 0$, a.e.$\mu \Rightarrow T^*f \geq 0$, a.e.$\nu$
$f \geq 0$, a.e.μ and $T^*f = 0$, a.e.$\nu \Rightarrow f = 0$, a.e.μ.
If h is a measurable function on $\mathscr{T}$, and $g = T^*f$,

$$\int_A hg d\nu = \int_{T^{-1}A} h(T) f d\mu$$

and so

(iii) $T^*[h(T) \cdot f] = h \cdot T^*f$. a.e.$\nu$.
It follows (see below) from (i), (ii), (iii) that T^* satisfies a Schwarz inequality

(iv) $[T^*(f_1 f_2)]^2 \leq T^* f_1^2 \cdot T^* f_2^2$ a.e.ν
with equality if and only if f_1/f_2 is a function of T a.e.μ.

$$|T^*f|^2 = \{T^*(\operatorname{sgn} f \cdot \sqrt{|f|} \cdot \sqrt{|f|})\}^2 \leq T^*|f| \cdot T^*|f|,$$

Therefore

(v) $|T^*f| \leq T^*|f|$ a.e.ν
with equality a.e.ν if and only if $\operatorname{sgn} f$ is a function of T, i.e. $T(x)$ constant $\Rightarrow \operatorname{sgn} f(x)$ constant.

$$\int |f_n - f| d\mu = \int T^*|f_n - f| d\nu \geq \int |T^*f_n - T^*f| d\nu.$$

Hence

(vi) $f_n \to f$ in mean $\Rightarrow T^*f_n \to T^*f$ in mean.

Proof. of (iv). Denote by λ_1, λ_2 real-valued measurable functions on $\mathscr{T}$. $\lambda_1(T), \lambda_2(T)$ are functions on $\mathscr{X}$.

$$T^*\{[\lambda_1(T) f_1 + \lambda_2(T) f_2]^2\} \geq 0 \qquad \text{a.e.}\nu,$$

$$\lambda_1^2 T^* f_1^2 + 2\lambda_1 \lambda_2 T^*(f_1 f_2) + \lambda_2^2 T^* f_2^2 \geq 0 \qquad \text{a.e.}\nu. \qquad [A.1]$$

First take λ_1, λ_2 as real constants. The set of points (exceptional points) at which [A.1] does not hold may vary with λ_1, λ_2; but for all rational values of λ_1, λ_2, the union E of exceptional points will have measure 0. Thus for all points in E^c, [A.1] is true for all rational λ_1, λ_2. Because of continuity, it is true in E^c for all real λ_1, λ_2, and so

$$[T^*(f_1 f_2)]^2 \le T^*f_1^2 \cdot T^*f_2^2. \qquad \text{a.e.}\nu.$$

If

$$[T^*(f_1 f_2)]^2 = T^*f_1^2 \cdot T^*f_2^2 \qquad \text{a.e.}\nu$$

$$T^*\{[\lambda_1(T)f_1 + \lambda_2(T)f_2]^2\} = [\lambda_1\sqrt{(T^*f_1^2)} + \lambda_2\sqrt{(T^*f_2^2)}]^2 \qquad \text{a.e.}\nu$$

for all functions λ_1, λ_2 on $\mathscr{T}$.
Take $\lambda_1 = -\sqrt{(T^*f_2^2)}$, $\lambda_2 = \sqrt{(T^*f_1^2)}$, then

$$T^*\{\lambda_1(T)f_1 + \lambda_2(T)f_2\}^2 = 0 \qquad \text{a.e.}\nu.$$

Therefore

$$\lambda_1(T)f_1 + \lambda_2(T)f_2 = 0 \qquad \text{a.e.}\mu$$

$$f_2/f_1 = -\lambda_1(T)/\lambda_2(T) \text{ is a function of } T \qquad \text{a.e.}\mu.$$

Conversely, if $f_2/f_1 = h(T)$ a.e.μ,

$$T^*(f_1 f_2) = T^*[h(T)f_1^2] = hT^*f_1^2 \qquad \text{a.e.}\nu$$
$$= T^*[f_2^2/h(T)] = h^{-1}T^*f_2^2 \qquad \text{a.e.}\nu$$

and

$$[T^*(f_1 f_2)]^2 = T^*f_1^2 \cdot T^*f_2^2 \qquad \text{a.e.}\nu.$$

3 L'Hôpital's rule. In Chapter 3 we use the following extension of the usual form of this rule.

Theorem

Let f, g be real-valued functions which are continuous in the open interval (a, b), with derivatives $f'(x)$, $g'(x)$ at each point x of (a, b). Suppose further that $g'(x) \neq 0$ in (a, b). If as $x \downarrow a$, $f(x)$ and $g(x)$ both $\to 0$, or if $g(x) \to \pm\infty$, then

$$\liminf_{x\downarrow a} \frac{f'(x)}{g'(x)} \le \liminf_{a\downarrow a} \frac{f(x)}{g(x)} \le \limsup_{x\downarrow a} \frac{f(x)}{g(x)} \le \limsup_{x\downarrow a} \frac{f'(x)}{g'(x)}.$$

Similar results hold for left-hand, and for two-sided limits.

Proof. Denote the four limits by l', l, L, L' respectively. We have to show that $l' \leq l, L \leq L'$. By Cauchy's formula, if $a < x < y < b$,

$$\frac{f(y)-f(x)}{g(y)-g(x)} = \frac{f'(\xi)}{g'(\xi)}, \text{ where } x < \xi < y.$$

Put

$$m(y) = \inf\{f'(x)/g'(x); a < x < y\},$$
$$M(y) = \sup\{f'(x)/g'(x); a < x < y\}.$$

We then have

$$m(y) \leq \frac{f(y)-f(x)}{g(y)-g(x)} \leq M(y). \qquad [A.2]$$

If $f(x), g(x)$ both $\to 0$ as $x \downarrow a$, this gives

$$m(y) \leq \frac{f(y)}{g(y)} \leq M(y).$$

Hence

$$l' = \lim_{y\downarrow a} m(y) \leq l, \quad L \leq \lim_{y\downarrow a} M(y) = L'.$$

We may rewrite [A.2]

$$m(y) \leq \frac{f(x)/g(x) - f(y)/g(x)}{1 - g(y)/g(x)} \leq M(y).$$

If $g(x) \to \pm\infty$ as $x \downarrow a$,

$$m(y) \leq \liminf_{x\downarrow a} \frac{f(x)/g(x) - f(y)/g(x)}{1 - g(y)/g(x)} = \liminf_{x\downarrow a} \frac{f(x)}{g(x)}$$

$$m(y) \leq l.$$

Hence $l' \leq l$. Similarly $L \leq L'$.
In both cases, when $\lim_{x\downarrow a} f'(x)/g'(x)$ exists

$$\lim_{x\downarrow a} f(x)/g(x) = \lim_{x\downarrow a} f'(x)/g'(x).$$

4 In Section 3 of Chapter 6 we require the following.

Theorem
If $\mathbf{A}, \mathbf{B}$ *are symmetric matrices of the same order* k, *and* $\mathbf{B}, \mathbf{A} - \mathbf{B}$

are both non-negative, then $|\mathbf{A}| \geq |\mathbf{B}|$, *and if* $|\mathbf{B}| > 0, |\mathbf{A}| = |\mathbf{B}|$ *if and only* $\mathbf{A} = \mathbf{B}$.

Proof. The result is obvious when $|\mathbf{B}| = 0$. Suppose $|\mathbf{B}| > 0$. $\mathbf{B}$ is then positive definite. First consider the case $\mathbf{B} = \mathbf{I}_k$, the identity matrix of order k. $\mathbf{A} - \mathbf{I}_k$ is then non-negative. Let λ be an eigenvalue of $\mathbf{A}$, and $\mathbf{V}$ a corresponding eigenvector.

$$\mathbf{AV} = \lambda \mathbf{V},$$
$$(\mathbf{A} - \mathbf{I}_k)\mathbf{V} = \lambda \mathbf{V} - \mathbf{V} = (\lambda - 1)\mathbf{V}.$$

Thus $\lambda - 1$ is an eigenvalue of $\mathbf{A} - \mathbf{I}_k$. Hence $\lambda - 1 \geq 0$, and so $\lambda \geq 1$. Therefore $|\mathbf{A}|$, the product of the eigenvalues of $\mathbf{A}$ is ≥ 1. If $|\mathbf{A}| = 1$, then every eigenvalue is 1, and so $\mathbf{A} = \mathbf{I}_k$.

In the general case $\mathbf{A} - \mathbf{B}$ is non-negative, and therefore $\mathbf{B}^{-1/2}(\mathbf{A} - \mathbf{B})\mathbf{B}^{-1/2} = \mathbf{B}^{-1/2}\mathbf{A}\mathbf{B}^{-1/2} - \mathbf{I}_k$ is non-negative. Hence $|\mathbf{B}^{-1/2}\mathbf{A}\mathbf{B}^{-1/2}| \geq 1$, i.e. $|\mathbf{A}||\mathbf{B}|^{-1} \geq 1$, with equality if and only if $\mathbf{B}^{-1/2}\mathbf{A}\mathbf{B}^{-1/2} = \mathbf{I}_k$, i.e. $\mathbf{A} = \mathbf{B}$.

5 Abel's binomial formula. We require this in Section 4 of Chapter 9. Perhaps the simplest and most easily remembered form of this is

$$(z + u)^n = \sum_{r=0}^{n} \binom{n}{r} u(u + r)^{r-1}(z - r)^{n-r} \qquad \text{[A.3]}$$

for positive integral n. Denote the right side by $f_n(z, u)$.

$$\begin{aligned}\frac{\partial f_n(z,u)}{\partial z} &= \sum_{r=0}^{n} \binom{n}{r} u(u + r)^{r-1}(n - r)(z - r)^{n-r-1} \\ &= n \sum_{r=0}^{n-1} \binom{n-1}{r} u(u + r)^{r-1}(z - r)^{n-1-r} \\ &= n f_{n-1}(z, u).\end{aligned}$$

Hence, if $f_{n-1}(z, u) = (z + u)^{n-1}$, then $f_n(z, u) = (z + u)^n + g(u)$. Putting $z = -u$, we have

$$\begin{aligned}g(u) &= f_n(-u, u) \\ &= \sum_{r=0}^{n} (-1)^{n-r} \binom{n}{r} u(u + r)^{n-1} \\ &= u\Delta_y^n (u + y)^{n-1} \text{ at } y = 0, \\ &= 0.\end{aligned}$$

Thus

$$f_{n-1}(z, u) = (z + u)^{n-1} \Rightarrow f_n(z, u) = (z + u)^n.$$

The statement [A.3] is true for $n = 1$, and therefore for all n. Interchanging r and $n - r$ in [A.3], we obtain the form required in Section 4 of Chapter 9

$$(z + u)^n = \sum_{r=0}^{n} \binom{n}{r} u(z - n + r)^r (u + n - r)^{n-r-1}. \quad \text{[A.4]}$$

□□□

REFERENCES

BILLINGSLEY, P. (1968) *Convergence of Probability Measures*. New York: Wiley.

DURBIN, J. (1968) The probability that the sample distribution function lies between two parallel straight lines. *Ann. Math. Statist.*, **39**, 398.

FISHER, R.A. (1925) Theory of statistical estimation, *Proc. Camb. Phil. Soc.*, **22**, 700.

HANNAN, J. (1960) Consistency of maximum likelihood estimation of discrete distributions. In *Contributions to Probability and Statistics. Essays in Honor of Harold Hotelling*, ed. Olkin, I. p. 249. Stanford: Stanford University Press.

KOLMOGOROV, A. (1933) Sulla determinazione empirica di una legge di distribuzione. *Inst. Ital. Attauari. Giorn.*, **4**, 1.

LE CAM, L. (1970) On the assumptions used to prove asymptotic normality of maximum likelihood estimates. *Ann. Math. Statist.*, **41**, 802.

LINDLEY, D.V. (1972) Review of *The Theory of Statistical Inference* by S. Zacks (Wiley), *J. Roy. Statist. Soc.*, **A 135**, 272.

PITMAN, E.J.G. (1965) Some remarks on statistical inference. In *Bernoulli, Bayes, Laplace*, ed. Neyman, J. & LeCam, L.M. p. 209. Berlin: Springer.

PRATT, J. (1960) On interchanging limits and integrals. *Ann. Math. Statist.*, **31**, 74.

ROYDEN, H. (1968) *Real Analysis*, (2nd edn.) p. 71. New York: MacMillan.

STECK, G.P. (1971) Rectangle probabilities for uniform order statistics and the probability that the empirical distribution function lies between two distribution functions. *Ann. Math. Statist.*, **42**, 1.

INDEX